INTRODUCTION TO CHEMISTRY LAB MANUAL

Donald Siegel

Rutgers, The State University of New Jersey—New Brunswick

Kendall Hunt
publishing company
4050 Westmark Drive • P O Box 1840 • Dubuque IA 52004-1840

Cover image © PhotoDisc, 2009

Kendall Hunt
publishing company

www.kendallhunt.com
Send all inquiries to:
4050 Westmark Drive
Dubuque, IA 52004-1840

Contents

Course Policies

REQUIRED ITEMS: BY WEEK 2

1. Make sure you are registered. Part of your termbill, if you registered for 171 was a $40.00 fee that covers the manual and the disposable items. If your name does not appear on our roster, you are not in the course and will not receive a manual, goggles or the rest of the chem kit.

2. Pad-lock for your drawer

DESCRIPTION OF COURSE

This course is a laboratory course. The grade your receive in based primarily on what you do during the labs. You are being graded on technique and accuracy. Occasionally, a student can do all the proper steps in the correct order, and still have bad results. This is usually because of poor technique, but regardless of the reason, we grade on the data you turn in.

The pre-lab assignment is due at the beginning of each lab. Immediately following is the quiz that covers the previous and current experimentation. All quizzes are closed-book. Students taking a quiz with a lab book or notebook open will have the quiz confiscated and receive a grade of zero on that quiz. **Latecomers will not get extra time or be allowed to make up the quiz, and *no credit* will be given for pre-lab assignment.** You will perform the experiment **after** your instructor discusses various aspects of the work, changes in procedure, and any tips. SAFETY GOGGLES MUST BE WORN PROPERLY IN THE LAB, if any one of the drawers is open. PENALTIES FOR VIOLATIONS WILL BE IMPOSED WITHOUT WARNING.

You will **not** need a separate lab notebook. Record all data directly on the data sheet provided and **show all calculations.** Any extra paper used must be handed in with the lab report. All entries have to be made in ink. A single line through the entry can be used to indicate any deletion—correctional fluid is not allowed. All lab reports must be submitted with the cover sheet at the end of the lab period whether the experiment is completed on the same day or carried over to the next laboratory period. ***NO EXTRA TIME WILL BE ALLOWED FOR UNFINISHED WORK*** and *no credit will be given if the report is taken out of the laboratory and submitted later.*

ABSENCE

An unexcused absence will result in a zero grade for the missed experiment; two (2) unexcused absences will constitute an automatic failing grade for the course. Anyone with more than 3 excused absences will be asked to drop the course and take it another time. Students with exactly 3 absences (of any sort) will be considered on a case-by-case basis. For excused absences, average of completed lab grades will be assigned.

To be excused from the lab or to arrange a make-up lab, you have to bring:

   i. a letter from your coach for **sports**

  ii. a note from your religious leader for **religious holidays** or

 iii. a doctor's note or your file number from the infirmary for **sickness**

A **Make Up Request Form** is required **at least 2 weeks in advance** for any absence other than illness. You must include all documentation with the request.

Due to the large size of this course, make-up labs may not be possible. Under extreme circumstances, a make-up lab may be possible in the same week with **prior consent from coordinators or stockroom personnel.** Only the coordinator(s) can approve absences for reasons other than illness. **Your TA cannot approve any absence.** If you missed a lab, contact coordinators or stockroom personnel as soon as possible to fill out a make-up authorization form. Without an authorization form, you will not be able to make up the lab. **Under no circumstances will a make-up lab be permitted once the last date for that experiment has past.** You must attempt to arrange a make-up for us to excuse the lab.

You must provide suitable documentation of all absences in a timely fashion. If you miss a lab and cannot make it up, fill out an **Excused Absence Form,** including written documentation. If we do not have written documentation from an acceptable source verifying that the absence was legitimate and excusable, within 15 business days from the absence, we will not count it. Absences are excused by the coordinators, not the TAs. If the coordinators have not received the documentation, then the absence will not be excused.

Missing the lab due to violation of the dress code is **NEVER** considered an excused absence!

BREAKAGE

When you break a piece of equipment, first clean up the broken glass (placing it in the appropriate disposal container), then go to the stockroom. You will fill out a breakage slip with the requested information including which item(s) you broke. The cost of these items will be totaled, and at the end of the semester the total will be submitted to the cashier's office. You will receive a bill for the broken items. **Do not pay the stockroom personnel for broken equipment. You will be billed at a later date.** Any equipment missing from your drawer at check-out will be added to this total. Failure to pay this bill promptly can result in deregristration, withholding of transcripts, and revocation of other privileges the University may deem appropriate. In addition, if you break an excessive amount of equipment, you will receive a grade of TF until the value of the breakage is paid.

In addition, failure to check out before the end of the semester will result in a fee being charged and a loss of points.

ACADEMIC HONESTY

You are being graded on the work *you* perform. Use of lab reports from other students (past or present) is expressly forbidden. Both the lender and the borrower are subject to severe penalties. Some discussion about the labs is acceptable at the discretion of the TA, but you must perform all the work (including the data analysis and answering of questions) yourself. The TA is free to ask you at any point to explain what you are doing. This is to help the TA instruct the confused and prevent copying of answers. If you are confused, ask for help. Don't just copy an answer.

Here are some common violations of the academic honesty policy and the penalties that have been assessed in the past:

Violation	Penalty
Possession of previous semester's lab report in class	Zero on lab
Manufacturing data	Zero on lab
Performing unauthorized experiments (horseplay)	Zero on lab
Copying quiz from neighbor	Zero on quiz
Second offence for any of the above	Failure of course

In addition to the penalties above, your academic dean will be informed of the infraction. You may be placed on academic probation, suspended, or expelled. Further, record of the violation can impact your ability to obtain professional credentials and/or licenses in the future.

CHAIN OF COMMAND

If you have a question about grading, you should first talk about it with your TA. If you are not satisfied with the explanation, you may raise the question with either of the course coordinators. We will not intervene for questions of a small number of points. Decisions made for safety (such as ejection for violation of safety rules) can be made by any TA, the stockroom personnel, or one of the coordinators, and these decisions are final and not subject to appeal. If you have a question about content, concepts or procedures then you may ask any TA or either of the coordinators for help. Use our office hours.

You must attempt to clear up any concerns you have about the grading of your reports as soon as possible. You have three weeks after the end of any given lab to request that its grading be reviewed. The coordinators will not consider appeals after the three weeks are up.

Frequently Asked Questions

Q. *I'm not feeling well and my lab's today. What should I do?*

A. Go to the Heath Center, or see your doctor. Most of the time, you can go to the Health Center that day and be seen by someone. Sometimes they are busy and will make an appointment with you. When you go there, leave your file open (waive your right to confidentiality). That way we can tell if you were really ill, or just didn't want to go to lab. Most of the time, students miss labs for minor illnesses (colds and flu) that the doctors can't really treat. We still need proof that you were sick.

With documentation, you can make up the lab. If making up the lab is impossible, we will average out the missing lab.

Q. *So, I have my doctor's note, how do I make up the lab?*

A. Each experiment is performed for only one week. Once the last section does that lab, it won't be done again. Here's the twist: we don't start on Monday. The semester starts on a Tuesday. So the week runs from Tuesday until the following Monday (in the fall, the Thanksgiving break changes the order: after Thanksgiving, the week runs from Thursday to the following Wednesday).

Once you have the note, call the number listed in your syllabus to arrange a make up. You can only make up the experiment on the same campus as you usually have lab. If there are no openings that fit in your schedule, then you cannot make up the lab. In that event, the lab will be excused and the points averaged out.

Q. *I have a Monday lab, so I can't make up the lab. What happens now?*

A. Students who cannot make up the lab with proper documentation are not penalized. We average out the missing lab. Since we drop the lowest quiz, the first lab you miss is the quiz that gets dropped.

Q. *I cannot come to lab because of a religious observance. What do I do?*

A. Get a letter from your religious leader stating that you are a member and will be observing that day. You must do this **at least 2 weeks before the lab you plan to miss!** I understand the Islamic holidays, such as the beginning of Ramadan, are determined by each celebrant by observing the moon. Nevertheless, you know when the holiday will begin, give or take a day. If the holiday might fall on your lab day, you still have to give 2 weeks notice.

Q. *I missed the lab because my car wouldn't start. What should I do?*

A. Nothing. We can't excuse students with transportation problems. Verifying the excuses is nearly impossible. There are only 2 exceptions: when a large number of students miss lab because the buses aren't running, we make an accommodation, and when you are involved in a car accident getting to the lab (provided you get a police report).

Q. *My family got non-refundable tickets for our vacation and the flight leaves the day of the final or the day I have lab. Can I arrange a make up?*

A. No. You must be in lab (or take the final) at your assigned time, unless you have a valid reason. Family vacations, gatherings, reunions are not acceptable reasons. Valid reasons include: officially (Rutgers) sponsored events (Rutgers sports teams, concerts, plays, etc.); death in the immediate family; illness or injury; car accident (with police report only) the day of the lab or exam.

Q. *I want to change sections because my friend is in another section or because I've heard that another TA is better or because I don't get along with my TA. How do I change sections?*

A. During the ADD/DROP period, go on-line or to the registrar's office and change sections. As long as there is room in the new section, you may move. We discourage this, but we cannot stop you. After ADD/DROP is over you may not change sections for any reason. If you don't like your TA and have tried everything to fix the situation, then you can either stick it out or drop the course and retake it some other semester.

If you are having trouble communicating with your TA, go to his office hours. If you are still not satisfied, see one of the coordinators.

Q. *I left my goggles at home. Can I borrow a pair from the stockroom?*

A. No. You may stay for the prelab quiz and talk and you may turn in your prelab. You may not stay in the lab without the proper attire, which includes safety goggles. You may go home and come back within the 3 hours that the lab runs. If you have to buy a new pair of shoes or goggles or pants, do it. We will not grant you a make up lab or excuse the absence.

Q. *I did everything the manual said, and I still got a low lab score. How is that fair?*

A. We test these experiments extensively. Most of the labs have been used here for more than 10 years. We look at the results of a large number of students and set the ranges of points based on what most students get. Students who use good technique get good results. Compared to most schools, our ranges are very generous. Most schools require that you get within 1% of the correct value to get full credit. On some of the labs, the range is ±20%. The labs are designed to emphasize lab technique, rather than understanding the theory behind it. You need the theory, but it is the physical application that matters.

Q. *I always run out of time. What should I do?*

A. You only get three hours each week to perform the experiment and complete the write up. Don't ask to stay an extra few minutes. It isn't fair to the other sections. Come to lab better prepared. Do more than one thing at a time. Work quickly and efficiently. For example, if you are waiting for a flask to cool or a chromatogram to run, do the post lab questions or write the equations or set up the next part of the experiment. Read the lab three times: once shortly after this week's lab, and second time when you start the prelab and the night before your lab. Sometimes, you can even answer some of the postlab questions before doing the experiment. Ask for help. Go to office hours.

Q. *I switched from 161 to 133 (or dropped 161 entirely). Should I also drop 171?*

A. That depends. It may be difficult for you to understand some of the concepts, but if you have been doing ok so far, you don't need to drop. However, most students who drop and retake with 161 usually get between $\frac{1}{2}$ and one grade better than those who continue in 171 before retaking 161.

Q. *I did the wrong prelab. Can I get more time to do the correct one?*

A. No. We give a copy of the syllabus to each student and post it on the web. You are responsible for preparing for the correct lab. If you did a prelab for an experiment that we will be doing later in the semester, hold on to it. You can still turn it in when it's due.

Q. *I have a class right before lab, so I usually arrive after the quiz has been given. Can I take the quiz at another time?*

A. No. You will have to choose which class is more important to you. If you couldn't change sections during ADD/DROP, then you should have dropped one of the classes and take it in a later semester.

Q. *I lost my manual. What should I do?*

A. Borrow from a friend and make copies of the required pages. We only order enough books for the anticipated enrollment. Occasionally, we have a few leftover because students dropped before picking up the manual. These can be purchased. See the stockroom attendant for details.

Q. *What should I know for the final?*

A. Everything. The final is based heavily on the prelabs, post labs and quizzes.

Q. *I have a common hourly exam during my lab. How do I arrange a make up lab?*

A. You don't. Regularly scheduled classes take priority over common hourly exams. The instructor of the course with the exam must make alternate arrangements for you. The following is from the University's policy on common hourly exams

---

Rules for Common Hour Examinations

The following rules shall govern the conduct of Common Hour Examinations and shall be enforced by the instructional deans of the faculties and colleges offering the courses in which Common Hour Examinations are given.

C. Certain scheduled Rutgers activities will take precedence over common hour examinations for students who are formally registered to participate in those activities. Activities that take precedence over common hour examinations include regularly scheduled Rutgers classes, scheduled Rutgers intercollegiate athletic practices, and scheduled Rutgers athletic events. Students who have conflicts between such activities and common hour examinations must be offered an alternate examination (see F., below).

F. If a student has a conflict between a common hour examination and a scheduled activity that takes precedence (see C.), then the department must offer the student an alternate examination within one week of the primary examination.

H. Alternate examinations must be offered at times that do not conflict with activities that take precedence.

J. Students and departments offering common hour examinations may disagree on whether a scheduling conflict constitutes grounds for missing a common hour examination. Student appeals in such cases will be referred to the dean of instruction in the student's college. The dean's decision will be binding on both the student and the department offering the common hour examinations. Student appeals must be submitted at least two weeks prior to the date of the conflict.

Q. *I have another exam at the same time as the lab final. What do I do?*

A. Fill out the **Conflict Exam Form.** There's one in each copy of the syllabus and posted on the website. Fill it out completely *before* the deadline (approximately 2 weeks before the exam). Fill out the form completely and return it as instructed. Giving it to your TA does not guarantee that it will get to the right place. If you are requesting the conflict due to two exams at the same time, three exams on the same day, or three exams in a row then you must attach a current copy of your registration showing which classes you are enrolled in. The official final exam times for all courses are assigned by the University (scheduling.rutgers.edu and click on the Final Exam link for the appropriate semester), and that's the list I go by. You do not need to send copies of the syllabus of the other courses or get letters from other instructors. You only need letters for other conflict reasons (religious holidays, etc.).

Q. *I am a special needs student who requires extra time for quizzes and exams. What should I do?*

A. You need documentation from your Dean's office. Once you have that we will make accommodations. We cannot give you extra time on the quizzes, but we will add in points to make up for it. We will make special arrangements for completing the calculations in the labs. For the final, fill out the **Conflict Exam Form** by the deadline just as though you had a conflict so that we can arrange for you to have extra time for the final.

Q. *I forgot to check when my other exams are and now it's after the deadline. What do I do?*

A. Hope that the other class is giving a conflict or make-up exam. We give you plenty of warning that you must check. It takes time to make all the arrangements and we pick a room and time based on the schedule of the students who qualified and the number of students who qualified. We are not going to make an exam just for you.

Q. *I have three exams in 24 hours, but they aren't in a row. Can I take the conflict exam?*

A. No. You must take the exam at the scheduled time. The relevant University Rules for what constitutes a conflict are listed below.

---

1. All final exams must be scheduled during the official Final Examination period as stated in the academic Calendar except those approved by the Dean of Instruction of the academic unit.

4. A student shall be said to have an exam conflict if that student has: 1) More than two (2) exams on one calendar day. 2) More than two (2) exams scheduled in consecutive periods (e.g. a student has exams scheduled for 4–7 PM and 8–11 PM on one day and 8–11 AM on the following day). 3) Two exams scheduled for the same exam period. Students having an exam conflict should contact the office of their college dean. Students having an illness requiring medical attention or conflict due to a religious observance should contact the instructor of the course(s) involved for information regarding arrangements for the make-up examination. The instructor may re-schedule the examination during the examination period and is responsible for re-scheduling, proctoring, and grading make-up examinations to accommodate students who have conflicts.

Q. *I forgot when the final was and so, I missed it. What do I do?*

A. Fail the course. We don't excuse students for forgetting when the final was. The final is a required part of the course and no one will pass who hasn't taken it.

Q. *My instructor moved the final in his course, so there's a conflict that won't show up on the official list. What do I do?*

A. Tell your instructor that he can't move the exam because the 171 instructor will not approve a conflict for such reasons. The University assigns the exam times to minimize conflicts, but some will occur anyway. When instructors change the times, it only creates more conflicts.

Q. *I know someone who got a better grade than I did but had fewer points. How is this fair?*

A. Because that student had a different TA. We correct for the different grading styles of each TA so that the final grade in the course reflects the level of understanding each student gains from it.

Lab Safety Rules

1. You are not permitted to be in the laboratory when a TA is not present.

2. Report all accidents and injuries, no matter how minor, to your TA.

3. You are only allowed to do authorized experiments.

4. Horseplay in the lab is unacceptable behavior and is cause for immediate ejection.

5. **You must wear safety goggles in the lab at all times.**

   - Contact lenses (hard or soft) are not permitted: trapped chemicals may cause injury to the eye.

   - Know the location and use of the closest eyewash, safety shower, and fire extinguisher.

   If you get chemicals in the eye, immediately flush the eye with copious amounts of water from the eyewash. For other parts of your body, wash the affected area thoroughly using the sink or safety shower.

6. Keep your book bags and other non-essential items at designated spaces only.

7. Bare feet, legs, or midriffs are not allowed in a chemistry lab. Sandals, open-toed or open backed shoes, shorts, or halters are not enough protection. If you have long hair it must be tied back. Old clothing or a laboratory apron or coat is highly recommended. **If you are not properly attired, you will not be admitted to the lab.** If you are ejected from the lab for improper dress, you will not be permitted back until you are properly dressed. If you miss the lab, or do not finish, you will not be permitted to make the lab up, and the absence will **NOT** be considered excused.

   ## IF YOUR BACK IS EXPOSED WHEN YOU BEND OVER THEN YOUR TOP IS TOO SHORT AND YOU WILL NOT BE ALLOWED TO WORK.

8. The vapors of a number of solutions are quite potent and can irritate or damage the mucous membranes of your nasal passages and throat. If you must smell a chemical, hold its container away from your face and waft its vapor gently toward your face with your hand. For reactions involving poisonous or noxious gases, use the hood by placing the container well within the marked lines. At Douglass, **ALL WORK MUST BE DONE INSIDE THE HOODS!**

9. Always keep burners under the hood. Never apply heat to the bottom of the test tube; always apply it to the point at which the solution is highest in the tube. A suddenly formed bubble of vapor may eject the hot and perhaps corrosive contents violently from the tube (an occurrence called "bumping").

10. No eating, drinking, or smoking in the lab. You may not bring in anything consumable, either.

11. Never taste chemicals or solutions—poisonous substances are not always so labeled.

12. Label all containers. Stock solutions must remain on the stock solution bench. Be sure to replace the same cap or stopper on the reagent bottles. Do not put medicine droppers or pipettes in the reagent bottles. Do not take too much stock solution. If you accidentally take more than you need, do not return the excess back in the reagent bottle; try to give it to another student or dispose of the excess as instructed. Grades may be reduced if instructions are not followed and materials are found where they should not be.

13. Although we do not provide gloves, you may wear them, if you choose to do so. Consult with your TA or the stockroom regarding the type of gloves you should consider. All experiments in this course can be safely performed without gloves.

14. Make sure your sink is cleaned out before leaving the lab.

15. Beware of hot glass—it looks cool long before it may be handled safely.

16. You must wash your hands at the end of lab even if you have been wearing gloves. This will prevent you carrying something out on your hands, which you later might get in your eyes or onto food.

17. **Inform your TA if you have a medical condition that requires special consideration.**

WHEN IN DOUBT, ASK YOUR INSTRUCTOR!

USE OF HOODS

We have two types of hoods that you will encounter: a low-flow laminar flow fume hood (called "traditional" hood from here on), and a canopy hood. Each is used in a different manner. You should understand both but know how to use the type of hood you have. Hoods are shared, so be courteous. Both types depend on air flow, so be careful not to block the vents.

CANOPY HOODS

These are older hoods, designed for student use. They are not designed to handle large amounts of very volatile compounds and provide no protection from spills or explosion. Essentially, they are air vents mounted over a portion of the work area. Whenever you are heating material or working with volatile compounds, you must use them.

To use them effectively, place the material inside the lines marked on your work area. The closer you get to the center of the hood, the more effective it is.

"TRADITIONAL" HOODS

These are newer hoods, and look like what most of us have come to think of as chemical hoods. They are metal boxes mounted on the benchtop. Our hoods have glass panels that slide left and right. In addition, the frame holding the panels slides up and down. They work by drawing air from the room into the box and out the top. They provide complete protection against exposure to volatile compounds under most conditions and, if used properly, provide some protection from spills, fire and explosions. You use them **WHENEVER** you are working with compounds. **ALL WORK SHOULD BE DONE INSIDE THE FUME HOODS.**

FIGURE 1

To use them, place the material inside the hood (see figure 1). You may slide the front up to do this if you need to. Most hoods require that the front be down most of the way. These hoods will operate with the front at any height; however they are designed to operate best with the front ALL the way down and the glass panels moved to allow access to the hood. Slide the two panels on your side of the hood one in front of the other so that both panes of glass are between you and your work. You can reach around the panels to handle the materials. This offers you the most protection against spills, splattering, fire and explosion.

Glassware and General Equipment

At Douglass, your drawer will be mostly empty. Most weeks, you will sign out a glassware kit. You have five minutes after you check it out to replace anything that's missing for free (the person who checked the kit out last will be charged). After that, anything that's missing will be charged to you. Your drawer will contain your goggles and disposable items (labels, filter paper, etc).

At Beck, on the Livingston campus, you will be assigned a locker. Each locker will contain beakers of various sizes ranging from 25 mL to 600 mL. These should be stored nested in the drawer. The graduations on the side of the beaker are very approximate (±20%).

You will have three Erlenmeyer (conical) flasks. These are useful for titrations. They are also graduated on the side, but again the volumes are very approximate. In addition, you will have a larger round flask, called a Florence flask.

You will have two graduated cylinders. These are calibrated more accurately.

You will have a round piece of glass called a watch glass and a thin walled porcelain cup called an evaporating dish.

You will have 2 sizes of test tube (some of you will have a third, very small size—we are no longer replacing those as they wear out). One of the large test tube will have no lip and be made of much thicker glass. It is called an ignition tube, because it can stand the forces for burning samples in it. The tubes can be nested in the test tubes rack.

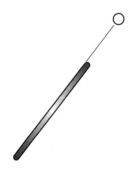

You will have medicine droppers, a funnel and a thermometer.

In addition to glass rods, called stirring rods, you will have a rod (either metal or glass) with a loop of metal wire in the end of it. This is called a Nichrome wire. The wire is a special alloy of nickel and chromium. When a material is placed on the loop and put in a flame, the material "burns" off emitting a color that can help to identify it.

You will have a crucible and lid for heating materials to a high temperature.

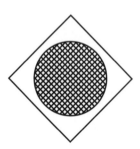

The wash bottle will usually contain deionized water. It is useful to transfer materials and to help in titrations. You will have another plastic bottle with a lid called a gas bottle.

The wire gauze is used to distribute heat from the microburner over the bottom of the material being heated.

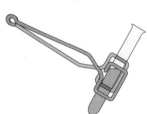

The crucible tongs are used to grab hot objects, like a crucible.

The test tube clamp allows you to hold a test tube when it's hot or when you don't want to spill the contents onto your skin.

The buret clamp (also called a utility clamp) will allow you to clamp items like burets, flasks, test tubes, etc. to a ring stand. It has adjustments to control the height on the stand, the pressure on the glassware, and the angle with the ring stand. Do not leave these out. They are expensive and you will be charged for it if you lose it.

The microburner is erroneously called a Bunsen burner. It is only a Bunsen burner with one particular arrangement of controls. The slits in the sleeve control the amount of air. The burn comes with a length of hose to attach it to a gas port.

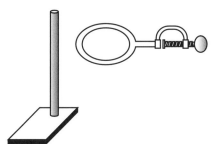

The ring stand and iron rings are among the common items you will find in the lab.

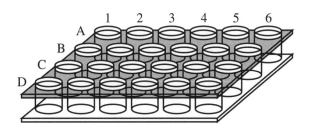

Your drawer will also contain a ruler, a 24-well plate, a clay triangle, brushes for cleaning the glassware, and a curved strip of metal called a scoopula that is used to scoop reagents out of bottles and beakers.

You will receive a number of disposable items when you get this book. They include gummed labels, filter paper, weighing paper, and most importantly, a pair of goggles. Do not share your goggles. There is a potential to transmit eye infections from one person to another. When you leave each week, put your goggles in your drawer. If you take your goggles to use in another class, you run the risk of leaving them home. We will not lend you goggles. At the end of the course, the goggles are yours to keep. You may need them again in future lab courses.

Good luck and enjoy!

Recording and Representing Data

One of the key elements of the scientific method is reproducibility. That both means in terms of the data we collect and the method we use to collect the data. The information we pass on to other scientists must be complete enough that they could repeat what we did exactly. Further, the conclusions we draw must be transparent. Our reports must be complete as well. Another scientist should be able to mentally perform our experiment, see our data and come to the same conclusions as we did.

That means that the integrity of our data is central to all science. In most labs, from industrial development labs to pharmaceutical productions plants to academic research labs to teaching labs, data is collected in a manner that permanently records it. Traditionally, that means in a lab notebook and in ink. In many commercial labs, such as in pharmaceutical companies, the chemist signs each page at the end of the day legally certifying that the data contained in the notebook is current and accurate. Many chemists have been fired because the data they collected either wasn't recorded in a timely fashion or not in a permanent method.

Mistakes happen when taking data. A famous chemist once said, "a neat notebook is the sign of a chemist that isn't working hard enough." We learn from our mistakes. Sometimes "mistakes" turn out to be correct. A good chemist will record over mistakes in such a manner that he can go back later in case the erroneous data turns out to be correct.

To save time, paper and expense, we have eliminated the traditional notebook. This lab manual *is* your notebook. Therefore, you should record all data directly in it and in ink. Correction fluid is expressly forbidden in the lab and especially over your data. You may use it to correct your answers on the prelab assignments, but not on the cover pages, data sheets or post labs.

Never trust your memory. Bring the data page with you to the balances or anywhere you go to record your data. Remember, you must use ink. There are two exceptions to this rule. Graphs are not the primary data, but a secondary representation of the data. Because we want crisp, sharp lines on graphs, we use very sharp pencils. See below for rules for graphing. The second exception occurs when the chemicals in use preclude the use of ink. The one example of that in this manual is chromatography. The solvents used in that lab would dissolve the pigments in ink, but not the graphite in pencils. Therefore, pencil is more permanent than ink on the chromatogram. Thus, on the chromatogram and *only* on the chromatogram, you should use pencil.

To correct mistakes, draw a straight line through the erroneous data and record the new data as near to the old as possible. See figure 1. Many labs have you initial next to the new data.

Should you forget, and record your data in pencil, pretend that you have a special pen that writes with pencil colored ink. We will take off points for writing in pencil. However, we will subtract **more** points for erasing and yet **more** points if you write over the pencil with pen. Just leave it alone, take your deduction and move on. No one has *ever* gotten a grade lower for writing in pencil.

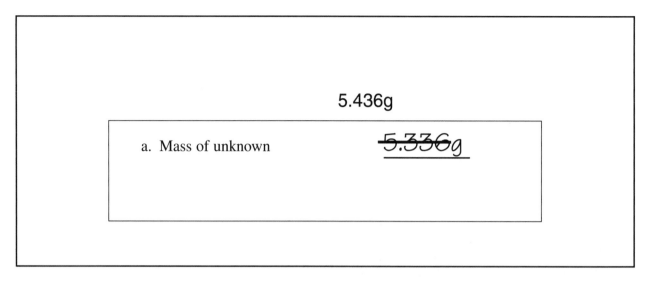

FIGURE 1. *Correcting data on the data sheet.*

SCIENTIFIC INTEGRITY

Your data represents you. If we can't trust your data, then we can't trust you. Some noted scientists have had their careers destroyed because they fabricated one unimportant experiment, or failed to note a colleague's contribution. Each lab is kind of a test and to some degree should be treated that way. Your work should be *your work*. Talk with your neighbors and, more importantly, with your instructor. Do not tell your labmates the answer. Show them how to get the answer.

Old labs are out there. Do not use them. If you are discovered using an old lab, not only will you be punished, but so will the student that you got the lab from. At least one student has been expelled for such behavior.

GRAPHING

Many times, we represent the data we collect in a visual manner. Charts and graphs make it easier for us to see relationships within our data. Because they re-present the data, that is present the data a second time, they do not have the same permanence required of them as the original data. They are part of the interpretation of the data. As such, we should **never** record data directly on a graph (exceptions include use of chart recorders and collection of spectra).

To be able to make inferences from graphed data, we need to be able to make good graphs. Here are a few general rules: graphs should be big—use 2/3^{rds} of the page for the graph (that includes the blank space for extrapolations); the bottom left hand corner need not be (0,0); every mark or line on a graph should be there for a reason—do not play connect-the-dots with the data. Random lines connecting the points don't convey useful information to the reader. Graphs should be well labeled. Include units and scales on your axes.

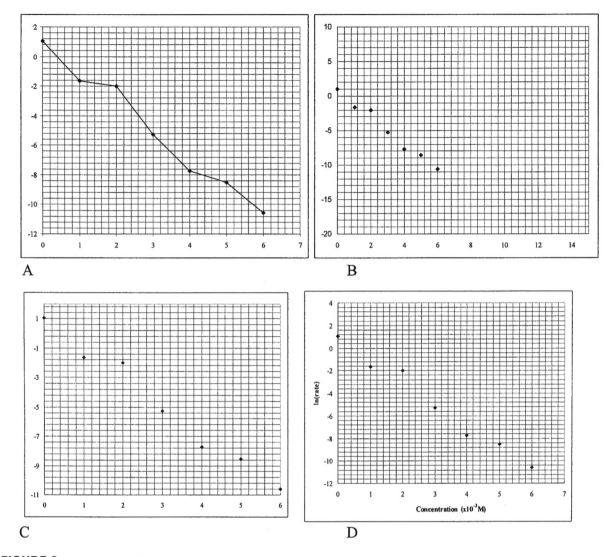

FIGURE 2. *Graphing data.* **A.** Poor—connect the dots. **B.** Poor—scale too big. **C.** Poor—scale too small. **D.** Good graphing technique.

Relationships within the data are frequently shown with a best fit straight line. Many math and statistics text can teach you linear regression. The truth is that the eye can see relationships better than most regression algorithms. In this course, you can draw the best line as determined by eye. Linear regression programs will give better estimates of the slope and intercept, but they will miss trends. A curve will often have the same r^2 value as randomly distributed data, but the eye will see the curve. In later courses you will learn the more rigorous analytical skills of fitting data to curves.

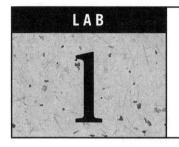

LAB

1

DENSITY: A PHYSICAL PROPERTY OF MATTER

INTRODUCTION

Density is a physical property. It represents the relationship between the mass of a substance and its volume. The density of any substance can be calculated given that both the mass of the substance and its volume are known. Mathematically, density is:

$$density = \frac{mass}{volume}$$

Density is commonly expressed in the following units for the following substances: Solids g/cm^3 or g/mL; Liquids g/mL or g/cm^3, Gases g/L. [Notice: 1 mL = 1 cm^3]

The mass of a substance can easily be determined using a balance, but the volume may or may not be easy to obtain. If one is working with a square or a rectangular shaped object, the volume may be calculated by taking the length times the width times the height (V = L × W × H). Many objects because of their irregular surfaces do not lend themselves to such an easy measurement. If an object is irregular, then a graduated cylinder or another piece of calibrated glassware may be used to determine the volume. Normally the volume of the object is found by the displacement of water. Water is added to a graduated cylinder and the volume is read. Let's say the volume reads 4.20 mL. The object is then added to the graduated cylinder and submerged under the water. Whereupon the volume of water in the graduated cylinder is read a second time. This time it reads 6.42 mL. The difference between the two volumes or readings is the volume of the object (6.42 mL – 4.20 mL = 2.22 mL = the volume of the object). This is called "determining the volume by difference."

A third method to determine the volume of an object is by Archimedes method. Objects in water are lighter (weigh less but have the same mass) than in air. That is because the displaced liquid pushes back on the object with a force equal to the mass of the liquid displaced. When an object floats, it displaces a volume of liquid with a mass equal to the mass of the object. A submerged object displaces a volume of liquid equal to the object's volume. The object will have an apparent loss of mass equal to the mass of the liquid displaced. Balances suitable for making these measurements are set up in the lab. See figure 1.

If the mass of the object is 2.00 grams and its volume is 2.22 mL, then its density can be calculated as follows:

Density = mass/volume = 2.00 g/2.22 mL = 0.901 g/mL

Adapted from *General Chemistry Laboratory Manual*. Fourth Edition by D.L. Stevens. Copyright © 2004 by Kendall/Hunt Publishing Company. Reprinted by permission.

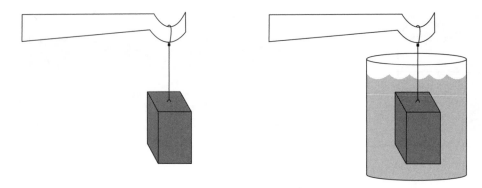

FIGURE 1. *The object weighed in air will weigh more than the object suspended in a liquid.*

In this experiment, you will be determining the density of the of an unknown metal. The mass will be determined by weighing it on an analytical balance (±1 mg or 0.001g). The density will be determined three ways: (a) by measuring the dimensions of the object; (b) by water displacement; and (c) by Archimedes method.

In making laboratory measurements, experimental errors occur because of the equipment being used or because of human error. This gives rise to accuracy and precision. Accuracy refers to how close the measurement is to the accepted value or theoretical value. For instance, the accepted value (density) of water is 1.00 grams/mL at 4.0°C. Errors relating to accuracy are called "percent error or percentage error." Percent error is calculated as follows:

$$\frac{|Actual\ value - theoretical\ value|}{Theoretical\ value} \times 100\%.$$

Precision refers to the repeatability of a measurement. For instance, if five different people were to weigh the same nickel on the same balance, would they all obtain the same value? The odds are they would not. They all may be close or they may obtain a wide range of measurements. Since the conditions were identical, except for five different people making each measurement, the results should be repeatable. Precision measurements give rise to an average, an average deviation, and a percentage or relative deviation.

A related term is sensitivity. This refers to the scale used to measure something. Some rulers read to the nearest 1/16th of an inch. A fine micrometer may read to the nearest 0.01 mm. There are costs and benefits to each tool. Generally, there is a trade off between ease of use and sensitivity. Highway scales do not need to weigh trucks to the nearest mg.

EXPERIMENTAL PROCEDURE

Obtain an unknown and record its unknown number on your data sheet. Determine its mass on an electronic analytical balance. Best practice is to determine it by difference. Weigh an empty beaker or watch glass. Place the metal in the beaker and weigh again. The difference will be your mass. This procedure avoids errors from miscalibrated and zeroed scales.

A. *Determining the Volume by Dimensions*

Measure all the dimensions of your unknown necessary to calculate its volume from the following formulae.

Volume of

$$\text{Rectangular parallelepiped} = \text{length} \times \text{width} \times \text{height}$$

$$\text{Cylinder} = \pi r^2 h$$

$$\text{Cylindrical tube} = \pi[r_{\text{outer}}^2 - r_{\text{inner}}^2]h$$

$$\text{Trapezoid (see figure 2)} = width \times height \times \frac{length\ 1 + length\ 2}{2}$$

B. *Determining Volume by Displacement*

Fill a graduated cylinder about half full with water. Use the smallest cylinder into which the object will fit. Read the volume in the cylinder and record it on your data sheet. To avoid breaking the cylinder, tip the cylinder at about a 45° angle and *gently* slide the object to the bottom of the cylinder. Be careful not to splash or spill any of the water. Return the cylinder to an upright position and record the volume on your data sheet. The difference will be the volume of your object.

C. *Determining Volume by Archimedes Method*

Tie a piece of string around the object. Leave a loop at the other end of the string. Hang the object on the support arm for the pan of a triple-beam balance. Record the mass to the nearest mg on your data sheet. Suspend the object in water by placing a beaker filled with water on the platform between the pan and the support arm. Record the temperature of the water. Make sure the object is completely immersed and is not touching the bottom or the sides of the beaker. Record the mass to the nearest mg on your data sheet. The difference in mass is the mass of water displaced. The volume of the object can be determined from the equation below and the density of water, found in table 1.

$$V_{\text{object}} = V_{\text{water displaced}} = \frac{\text{decrease in mass of object}}{\text{density of water}}$$

Record your results on the data sheet.

D. *Calculations and Interpretation*

Calculate the volume of the object from the three sets of data you collected: by dimensions, by displacement, and by Archimedes Method. Take the mass recorded on the analytical balance (line "a" on the data sheet) and divide it by each of the three volumes you determined. Notice that your densities

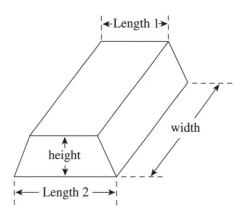

FIGURE 2. *A trapezoid.*

TABLE 1
DENSITY OF WATER

Temp (°C)	Density (g/mL)
17	0.9988
18	0.9986
19	0.9984
20	0.9982
21	0.9980
22	0.9978
23	0.9975
24	0.9973
25	0.9971
26	0.9968
27	0.9957

will vary from one method to the next. Based on the experiments themselves (your feel for how likely they are to give an accurate reading), pick one of the three densities as your "most trusted" value. For example, if the object was a cylindrical rod, but the cross section wasn't a perfect circle, you wouldn't pick "by dimensions" as most trusted. The number of significant figures you can report is usually a good estimate of how well you can trust a number, but not the only one. You did the experiment, and you know what went wrong. Take your most trusted value and compare it to the values in Table 2. Identify the material that your unknown is made of as the material with the density closest to your most trusted value. Write the identity on your cover sheet along with the density you found in Table 2. Find the % error between this value and the density you measure that is closest to this value. Do not be alarmed if the % error is large.

TABLE 2
DENSITIES OF VARIOUS MATERIALS

Metal	Density (g/cm^3)
Aluminum	2.70
Brass	8.90
Chromium	7.19
Copper	8.93
Iron	7.8
Lead	11.35
Magnesium	1.74
Molybdenum	10.22
Steel	7.76
Tin	7.28
Titanium	4.54
Tungsten	19.35
Zinc	7.14
Zirconium	6.50

LAB 2

PAPER CHROMATOGRAPHY

INTRODUCTION

Just after the turn of the 20th century, a Russian botanist named Mikail S. Tsweet developed the chromatography process. Tsweet originally used the process to separate plant pigments. "Chromo" means color and "graphic" means to write. Chromatography thus means color writing.

There are four main types of chromatography:

a. column chromatography

b. gas chromatography

c. paper chromatography

d. thin-layer chromatography

The mechanics and processes for all four types of chromatography follow the same general principles. A mixture of solutes in a *mobile phase* pass through or over a selective adsorbing medium, the *stationary phase.* As the mobile phase migrates through the stationary phase, separation occurs because the solutes have different affinities for the two-phased system (the mobile and the stationary parts). This difference in affinities results in the separation of the solutes.

In column chromatography, the stationary phase is made of small particles packed in a glass tube. As the mobile phase (a gas or a liquid) migrates through the particles, they are separated by their interaction with either the stationary phase or the mobile phase or both.

In gas chromatography, the stationary phase is made of small particles packed in a glass tube. The mobile phase is usually an inert gas like helium. The mixture to be separated is carried through the stationary phase by the inert gas. Gas chromatography is popular with organic chemists as it is a very effective means of separating and identifying organic compounds.

In paper chromatography, the paper is the stationary phase and the liquid or developing solution is the mobile phase.

Thin-layer chromatography is very similar to paper chromatography but it uses a dried gel or slurry laid down on a smooth surface like a square of glass. The glass and the slurry are the stationary phase. The developing solvent is the mobile phase.

Adapted from *General Chemistry Laboratory Manual.* Fourth Edition by D.L. Stevens. Copyright © 2004 by Kendall/Hunt Publishing Company. Reprinted by permission.

In this experiment, you will run two paper chromatograms. The first chromatogram will be used to find where various metal ions run in this solvent system. The paper represents the stationary phase and the developing solution or eluting solution (in this case, an HCl:acetone mixture) represents the mobile phase of the system. Some ions will migrate with the mobile phase and some will adhere to the stationary phase. It is these interactions that create the separation of the ions. Along with the known ions, you will be provided with two unknown solutions (A and B) containing ions.

As the ions move through the paper via capillary action, they will migrate at different rates due to their attractive forces (e.g., dipole-dipole, ion-dipole, dipole-hydrogen bonding, etc. interactions) for the paper or the stationary phase. This interaction between the mobile and the stationary phase will cause the dyes to separate into their individual components. This interaction between the mobile and stationary phases determines the rate at which the separation occurs. A variety of developing solutions have been tested to determine which developing solution provides the best separation. A developing solution can consist of a water-organic mixture, a water-salt mixture, a water-organic-salt mixture, etc. Many have been created and tested. Cost, speed, and efficiency determine which developing solution is used. The particular mobile phase you will be using is a mixture of 6M HCl and acetone (CH_3COCH_3). Acetone is less polar than water, so the mobile phase is less polar than aqueous HCl. The paper is made of cellulose which has a surface covered in polar –OH groups.

Transition metals in solution form complexes. In water they form complexes with water: $[M(H_2O)_x]^{2+}$ for metals in the +2 state, and $[M(H_2O)_x]^{3+}$ for metals in the +3 state. Different metals complex with different numbers of water. As the chloride concentration increases, it replaces the water to form the complex $[M(H_2O)_xCl_y]^{(2-y)}$, in the case of a +2 metal. The more highly charged the complex, the greater the affinity for the more polar stationary phase and the slower the ions move.

The second chromatogram will be run using your unknown solutions each containing one or more metal ions. The four metal ions are cobalt(II), Co^{2+}; copper(II), Cu^{2+}; iron(III), Fe^{3+}; and nickel, Ni^{2+}. All of the metal ions cannot be identified by their colors. Because some the metal ions are not visible to the human eye, they will have to be "developed." Several of the metal ions will have to be reacted with a special reagent to make the metal ions visible to the human eye. To determine which reagent to use, before running the first chromatogram you will run a spot test to see how each ion behaves with the various developing reagents.

Once a chromatogram has been developed, it needs to be interpreted. Normally this is reported as the Rf value (retention or retardation factor). Mathematically it is expressed as:

$$R_f = \frac{Distance\ traveled\ by\ ion}{distance\ traveled\ by\ solvent\ front}.$$ See figure 1.

If the conditions of a run are relatively constant, the R_f value should be relatively constant. Factors that can affect the R_f value on a chromatogram are: large temperature changes, variations in the paper itself, overlapping paper edges, the developing solvent used, etc.

Chromatography can be used to identify compounds and ions and to separate components of a mixture. The degree to which species are separated is call *resolution*. See figure 2. Because R_f is a ratio, it is independent of resolution.

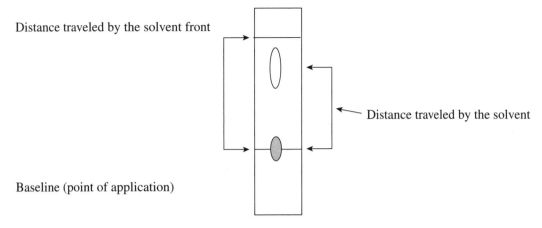

Distance traveled by the solvent front

Distance traveled by the solvent

Baseline (point of application)

FIGURE 1. *How to interpret a chromatogram*

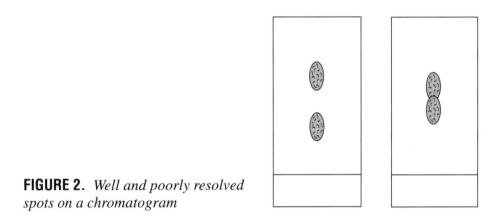

FIGURE 2. *Well and poorly resolved spots on a chromatogram*

LABORATORY SAFETY

1. Departmental approved safety goggles worn over your eyes and appropriate dress code must be adhered to at all times while in the laboratory.

2. Avoid getting the dyes on your skin and clothes, as they will leave stains. If you do get them on your skin or clothes, use water to remove them.

3. The mobile phase for the chromatograms is a mixture of 10% 6M hydrochloric acid, HCl, in 90% acetone, CH_3COCH_3. Acetone is commonly used in fingernail polish remover. The HCl can cause burns and skin irritation. The acetone vapors may be irritating to the lungs. Avoid both contact and inhalation of this developing solution.

4. *Concentrated ammonia, NH_3, must be used in the fumehood because it is a very strong lung irritant and should not be inhaled!*

5. The developing reagents include 1.0% dimethylglyoxime solution. As with all chemicals, avoid physical contact. If you do get dimethylglyoxime on your skin, wash it off with soap and large amounts of water.

PROCEDURE

I. Spot Test

Take three pieces of filter paper out of your chem. kit. Fold them into quarters. Spot a small amount of the four ions, one in the center of each quarter. Spread one of the developing reagent on the spots on each filter paper. In the case of dimethylglyoxime, the paper must first be exposed to ammonia vapors. It is only a basic dimethylglyoxime-ion complex that shows color. Record your results on the data sheet and decide which reagent to use to visualize each ion.

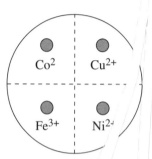

FIGURE 3. *Spot test.*

II. Preparing a Chromatogram

A. Obtain two pieces of chromatographic paper. On one piece of the chromatographic paper, draw a line parallel to the long dimension of the paper using a pencil (not ink). This line should be about 1.5 cm from the bottom of the paper. This is your baseline.

B. Fold the paper into fourths like an accordion with the folds perpendicular to the baseline (see figure 4). Using a pencil, put hash marks across the baseline in the middle of each panel and label under them "Co," "Cu," "Fe," and "Ni." The ions can go in any order. Use the intersection of the hash mark and the baseline as the target when you spot the solutions.

C. A capillary tube will be used to place the knowns and unknowns on the chromatogram paper. When you apply the solutions do not use large amounts. Do not intermix the capillaries or you will contaminate the samples. You will be given a single capillary tube for both your unknowns. Use a different end for each solution, or break the tube in half, carefully. To each spot apply two applications, one over the other, allowing the previous sample to dry before the next is applied. The chromatogram may be waved in the air to dry the spots.

III. Developing Your Paper Chromatogram

D. Pour approximately 5 mL of mobile phase into a 400-mL beaker. Cover the beaker with a watch glass while you are spotting the chromatogram. Place the chromatogram paper in the beaker. Try to

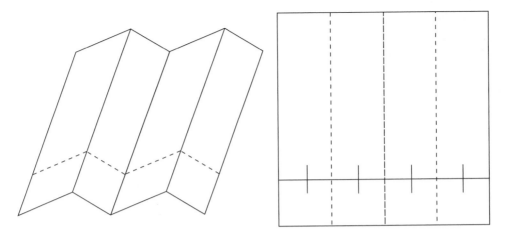

FIGURE 4. *Preparing a chromatogram.*

avoid having your chromatogram touch the walls of the beaker. While your chromatogram is in the beaker, *notice the location of the baseline at the bottom of the chromatogram. Your developing solution must not be equal to or be above the baseline!* If the developing solution is at or above the baseline, your spots will go into solution and will not migrate up the chromatogram. If this occurs, you will have to prepare a new chromatogram.

E. With the proper amount of developing solution added to the beaker, place your prepared chromatogram into the beaker. Cover the beaker with a watch glass. The chromatogram is sealed in the beaker and this allows the atmosphere in the beaker to become saturated with the vapors of the developing solution. This establishes an equilibrium system. Allow the developing solution to rise within 2 to 3 cm of the top of the paper. This may take 15–20 minutes. When the developing solution is 2 to 3 cm from the top of the chromatogram (*this is your solvent front*), remove the chromatogram and place it flat on a paper towel. *Immediately and with a pencil,* mark the solvent front—this is the distance the developing solvent has moved up the chromatogram. Your solvent front will be visibly wet. This must be done immediately as the solvent front will continue to migrate while the chromatogram is drying.

F. Allow the chromatogram to dry. Once the chromatogram is dry, apply the developing reagent you picked from the spot test to locate the ions on the chromatogram. With the glass tube provided streak the solution from the solvent front down towards the baseline. Using a pencil, *circle the outermost outline of each color that is visible.* Those metal ions that are visible will have only one color present. In the very the center of each visible color place a pencil dot. This center dot represents an average of the movement of the metal ion. Remember, the dimethylglyoxime-ion complex may not be visible until it has been exposed to ammonia vapors. Allow the chromatogram to dry again and hold it over the concentrated ammonia and circle any new spots that appear. Remember: All of the metal ions are NOT VISIBLE to the naked eye. Those metal ions that are not visible to the naked eye will have to be "*developed*" [made visible] by reacting the invisible metal ion with a reagent which makes the metal ion visible. Occasionally the colors of the unknown metal ions will overlap and thus your pencil circles will also overlap. Measure the distance traveled by each spot along with the distance traveled by the solvent front and record these measurements on your data sheet (**in ink!**).

IV. Interpreting Your Paper Chromatogram

G. Calculate the R_f value for each known metal ion. Record these values in the data table. Each metal ion will have its own R_f value. You may want to set up and run your second chromatogram before doing these calculations.

V. Identifying Your Unknown

H. Set up another piece of chromatographic paper as you did for the first run. The four lanes for this chromatogram will be:

Lane	Solution
1	Unknown A
2	Mixture of all 4 ions (known)
3	Unknown B
4	Mixture of all 4 ions (known)

Run the chromatogram in the same mobile phase as the first run. Try to keep the spots small. Use the first chromatogram as a guide to decide if you need more or less solution on the chromatographic paper. If the level has dropped too much, you may have to add some more running solution to the beaker. Remember, the level **MUST NOT** be above the baseline of your chromatogram. When the solvent front has risen to 2–3 cm from the top, remove the paper from the beaker and mark the solvent front. Develop the ions with the same solutions as before. This time, you know approximately where to look for the ions. Streak *across* the paper at the appropriate level of the ion you are attempting to make visible. You many have more than one spot per sample for the unknowns, and the knowns will have all 4 ions. Determine the R_f values for all the ions you detect and record them on your data sheet. Based upon your R_f values and by visual examination, determine which metal ions were present in your two unknowns, A and B.

I. Staple both of your paper chromatograms and your spot tests to the front of this laboratory report before handing it in.

PRE-LABORATORY QUESTIONS

1. Which is the *stationary phase* in the this experiment?

2. Which is the *mobile phase* in the this experiment?

3. Why do the metal ions separate from each other as they migrate through the paper chromatogram?

4. Which reagent is the best to develop Mn^{2+}? Briefly explain.

$Mn^{2+}(aq) + 2OH^-(aq) \rightarrow Mn(OH)_2$ (s, pale pink)

$Mn^{2+}(aq) + C_2O_4^{2-}(aq) \rightarrow MnC_2O_4$ (s, pale pink)

$2Mn^{2+}(aq) + [Fe(CN)_6]^{4-}(aq) \rightarrow Mn_2Fe(CN)_6$ (s, pale green)

$2Mn^{2+}(aq) + 5BiO_3^-(aq) + 14H^+(aq) \rightarrow 2MnO_4^-$ (aq, deep purple) $+ 5Bi^{3+} + 7H_2O(l)$

5. Why don't you use pen on the chromatogram?

NAME

COVER SHEET

PAPER CHROMATOGRAPHY

Purpose: To separate four transition metal ions by paper chromatography and identify the components of a mixture.

Procedure: The procedure for this experiment was as in the write up except (list any changes in procedure):

Results:

Knowns (Report a representative R_f value for each ion)

Ion	R_f	Reagent used
Co^{2+}	_____	_____
Cu^{2+}	_____	_____
Fe^{3+}	_____	_____
Ni^{2+}	_____	_____

Unknowns

Unknown #	Ion	R_f
_____	_____	_____
_____	_____	_____
_____	_____	_____
_____	_____	_____
_____	_____	_____
_____	_____	_____
_____	_____	_____
_____	_____	_____

Conclusions and Comments:

DATA SHEET 1

Spot Tests

Report color observed with each developing reagent.

Ion	$K_4Fe(CN)_6$	KSCN/Acetone	Dimethylglyoxime/NH_3
Co^{2+}	_____	_____	_____
Cu^{2+}	_____	_____	_____
Fe^{3+}	_____	_____	_____
Ni^{2+}	_____	_____	_____

Which reagent will you use to develop for each ion?

Co^{2+} _____ Cu^{2+} _____

Fe^{3+} _____ Ni^{2+} _____

You may check with your instructor for advice about selecting reagents.

Chromatogram 1

Distance traveled by solvent front: _____

Ion	Distance to Center of Spot	R_f value
Co^{2+}	_____	_____
Cu^{2+}	_____	_____
Fe^{3+}	_____	_____
Ni^{2+}	_____	_____

DATA SHEET 2

Chromatogram 2

Distance traveled by solvent front: _____

Lane 1:

Ion	Distance	R_f
Co^{2+}	_____	_____
Cu^{2+}	_____	_____
Fe^{3+}	_____	_____
Ni^{2+}	_____	_____

Lane 2: (You should have at least one spot and may have up to 4)

Spot	Distance	R_f	Identity
1	_____	_____	_____
2	_____	_____	_____
3	_____	_____	_____
4	_____	_____	_____

Lane 3:

Ion	Distance	R_f
Co^{2+}	_____	_____
Cu^{2+}	_____	_____
Fe^{3+}	_____	_____
Ni^{2+}	_____	_____

Lane 4: (You should have at least one spot and may have up to 4)

Spot	Distance	R_f	Identity
1	_____	_____	_____
2	_____	_____	_____
3	_____	_____	_____
4	_____	_____	_____

Compare the R_f values and explain why you have identified the ions as you have.

POST-LABORATORY QUESTIONS

1. Explain why the R_f values of the ions on the same chromatogram were not all identical.

2. Having now performed two chromatogram experiments, suggest two possible applications for paper chromatography other than the identification of metal ions. Admittedly your experience is limited here, but let your imagination go.

3. During a chromatography experiment, a student calculated two R_f values. They are: A: 0.62 and B: 0.30. The solvent front for the original chromatogram was 12.0 cm. If A and B were run on a second chromatogram and the solvent front was 25.0 cm, how far apart would A be from B?

4. What will the effect be on the resolution and R_f values in the following situations?

   a. The watch glass is left off the beaker during the run.

   b. The chromatogram was left in the beaker too long so that the solvent front went all the way to the top of the paper 5 minutes before the paper was removed.

   c. The spots were too big.

   d. A student used a shorter piece of chromatographic paper.

NET IONIC EQUATIONS

BACKGROUND INFORMATION

When metal and non-metal atoms join, they usually form an ionic compound. Ionic compounds contain charged ions rather than neutral atoms. Ions are chemical species that have either a positive or negative charge. A positively charged ion, or cation, has fewer electrons than protons. A negatively charged ion, or anion, has more electrons than protons. Metal atoms usually form cations, and non-metal atoms tend to form anions. Many anions are polyatomic: they are composed of several non-metal atoms bonded together as ions. Anions and cations are attracted to each other by their opposing electrical charges. Sodium chloride is an ionic compound formed from a sodium cation (Na^+) and a chloride anion (Cl^-). The force holding the compound together is electrostatic attraction, and is governed by Coulomb's Law (Eq. 1.)

$$F \propto \frac{q_1 q_2}{r_{12}} \qquad \text{(Eq. 1)}$$

Table 1 lists the names and formulae of some common cations and anions.

Writing the Formula of an Ionic Compound

Ionic compounds, like all chemical compounds, are electrically neutral substances. Therefore, cations and anions in an ionic compound must be combined in ratios such that the net charge on the compound is zero. Because the charge on each Na^+ ion is +1, and the charge on each Cl^- ion is –1, these ions combine in a 1:1 ratio to form NaCl. Similarly, each Ba^{2+} ion requires two Cl^- ions to counterbalance its +2 charge. Hence the formula of barium chloride is $BaCl_2$. Note that the number of Cl^- ions present for every Ba^{2+} ion is indicated by the subscript following the Cl. Also note that we do not indicate the charges in the formula.

With complex ions, that is ions that contain more than one atom such as hydroxide, we cannot write the barium hydroxide formula as $BaOH_2$, because this expression indicates a compound composed of one Ba, one O, and two H atoms. To indicate two OH^- ions, we put parentheses around the OH, and write a subscript 2 outside the parentheses: $Ba(OH)_2$. This use of parentheses with a subscript 2 means "two of everything inside the parentheses." If the subscript for a polyatomic ion is 1, there is no need to use parentheses in the chemical formula. Hence we write the chemical formula for sodium hydroxide as NaOH, not Na(OH).

TABLE 1
COMMON IONS AND THEIR CHARGES

Cations (+)		Anions (−)	
Monatomic	*Polyatomic*	*Monatomic*	*Polyatomic*
1+ Hydrogen, H^+ Cesium, Cs^+ Copper (I), Cu^+ Lithium, Li^+ Mercury (I), Hg^{2+} Potassium, K^+ Silver (I), Ag^+ Sodium, Na^+	**1+** Ammonium, NH_4^+	**1−** Bromide, Br^- Chloride, Cl^- Fluoride, F^- Iodide, I^-	**1−** Acetate, $C_2H_3O_2^-$ (CH_3COO^-) Bicarbonate, HCO_3^- Chlorate, ClO_3^- Cyanide, CN^- Hydrogen sulfate, HSO_4^- Hydroxide, OH^- Nitrate, NO_3^- Nitrite, NO_2^- Perchlorate, ClO_4^- Permanganate, MnO_4^- Thiocyante, SCN^-
2+ Barium, Ba^{2+} Beryllium, Be^{2+} Cadmium (II), Cd^{2+} Calcium, Ca^{2+} Chromium (II), Cr^{2+} Cobalt (II), Co^{2+} Copper (II), Cu^{2+} Iron (II), Fe^{2+} Lead (II), Pb^{2+} Magnesium, Mg^{2+} Manganese (II), Mn^{2+} Mercury (II), Hg^{2+} Nickel (II), Ni^{2+} Strontium, Sr^{2+} Tin (II), Sn^{2+} Zinc (II), Zn^{2+}	**2+** Mercury (l), Hg_2^{2+}	**2−** Oxide, O^{2-} Peroxide, $O2^{2-}$ Sulfide, S^{2-}	**2−** Carbonate, CO_3^{2-} Chromate, CrO_4^{2-} Dichromate, $Cr_2O_7^{2-}$ Hydrogen Phosphate, HPO_4^{2-} Sulfate, SO_4^{2-} Sulfite, SO_3^{2-}
3+ Aluminum, Al^{3+} Cobalt (III), Co^{3+} Chromium (III), Cr^{3+} Iron (III), Fe^{3+}	**3+**	**3−** Nitride, N^{3-}	**3−** Arsenate, AsO_4^{3-} Ferricyanide, $Fe(CN)_6^{3-}$ Phosphate, PO_4^{3-}

Dissolving Ionic Compounds

Water is a polar solvent with the ability to disrupt the electrostatic interactions within many ionic solids. For such solids, the ion-water interactions are strong enough to cause the dissolution of the solid. When $NaCl$ dissolves in water, the separated sodium and chloride ions are hydrated, as shown in Equation 2.

$$NaCl(s) \xrightarrow{H_2O} Na^+(aq) + Cl^-(aq) \qquad \text{(Eq. 2)}$$

When we write equations, we should indicate the states of the reactants and products. To do this, we use the notations (s) for solid, (*l*) for liquid, (g) for gas, and (aq) for a substance dissolved in water. More commonly, we write this as in Equation 3:

$$NaCl(s) \rightarrow NaCl(aq) \qquad \text{(Eq. 3)}$$

Calcium chloride, $CaCl_2$, is another water-soluble ionic compound. We can describe the dissolution of $CaCl_2$ by Equation 4, where we have also indicated the colors of the reactant and product. This is not the normal way to write reactions. Usually, we omit the colors.

$$CaCl_2(s, white) \rightarrow CaCl_2(aq, colorless) \qquad \text{(Eq. 4)}$$

When a $CaCl_2$ solution is mixed with a $AgNO_3$ solution, a white solid appears in the mixture, as shown in Equation 5.

$$CaCl_2(aq, colorless) + 2\,AgNO_3(aq, colorless) \rightarrow 2AgCl(s, white) + Ca(NO_3)_2(aq, colorless) \qquad \text{(Eq. 5)}$$

The solid, called a precipitate, is silver chloride ($AgCl$), an ionic compound that is not water soluble at 20°C.

If we show the individual hydrated ions in equation 5 as separate entities and the precipitate as a species of its own, we get Equation 6.

$$Ca^{2+}(aq) + 2Cl^-(aq) + 2Ag^+(aq) + 2NO_3^-(aq) \rightarrow 2AgCl(s) + Ca^+(aq) + 2NO_3^-(aq) \qquad \text{(Eq. 6)}$$

We call such an equation the **complete ionic equation** for the reaction. Notice that the same number of Ca^{+2} ions and NO_3^- ions appear on both sides of equation 6. Because these ions are not actively involved in the reaction, we refer to them as **spectator ions.** We can eliminate the spectator ions from equation 6 to obtain the **net ionic equation** for the reaction:

$$Ag^+(aq) + Cl^-(aq) \rightarrow AgCl(s, white) \qquad \text{(Eq. 7)}$$

Note that the stoichiometric coefficients of "2" have been eliminated as well, because they are not needed.

The ions in solution are free to move around. By random chance, they collide with other ions. If an ion encounters another an that makes an insoluble pair (see the guidelines below), a precipitation occurs. If all possible cation-anion combinations are soluble in the solution solvent, we say that no reaction occurs, because all the ions remain solvated. When we mix aqueous solutions of AX and of BY, the solubility of AY and BX will determine whether or not precipitation occurs.

The solubility of a particular ionic compound in water depends on both the identity and relative concentrations of the ions present. The relative concentrations comes into play through equilibrium, which is beyond the scope of this experiment. Hence solubility must be experimentally determined. Testing of many different ionic compounds has led to the establishment of some general solubility guidelines for cation-anion combinations. We use the following guidelines to determine the solubility of new ionic combinations. To be useful, the guidelines must be applied in the order presented.

1. Most compounds containing Na^+, K^+, or NH_4^+ ions are soluble.

2. Most compounds containing NO_3^-, CH_3COO^-, or ClO_4^- ions are soluble.

3. Most compounds containing Ag^+, Pb^{2+}, or Hg_2^{2+} ions are insoluble.

4. Most compounds containing Cl^-, Br^-, or I^- ions are soluble.

5. Most compounds containing CO_3^{2-}, S^{2-}, or OH^- ions are insoluble.

6. Most compounds containing SO_4^{2-} ions are soluble; however, $BaSO_4$ and $CaSO_4$ are insoluble.

7. Most compounds containing IO_3^- are soluble; however, $Ba(IO_3)_2$ is insoluble.

Establishing the Solubility of Some Precipitates in CH₃COOH or HNO₃ Solution

If we mix aqueous solutions of $Zn(NO_3)_2$ and NaOH, a white precipitate forms. Based on solubility guidelines 1 and 2, we can predict that the precipitate is $Zn(OH)_2$, not $NaNO_3$. We write the equation for the reaction of $Zn(NO_3)_2$ solution with NaOH solution as:

$$Zn(NO_3)_2(aq) + 2NaOH(aq) \rightarrow Zn(OH)_2(s) + 2NaNO_3(aq) \qquad \text{(Eq. 8)}$$

Acids in aqueous solution produce $H^+(aq)$ ions. A few of the many different acids are listed below. The hydrogen ion lost from each acid molecule is indicated in boldface.

acetic acid	**CH₃COOH**	nitric acid	**HNO₃**
hydrochloric acid	**HCl**	sulfuric acid	**H₂SO₄**

Note that the anion remaining after the H^+ ion has been lost is in Table 1. CH_3COOH and HNO_3 solutions are useful sources of H^+ ion because most compounds containing CH_3COO^- or NO_3^- ions are soluble in water (guideline 2), while some compounds containing Cl^- or SO_4^{2-} ions are insoluble in water.

When we add acetic acid solution (CH_3COOH) to an aqueous solution containing a $Zn(OH)_2$ precipitate, the solid dissolves. The equation for this dissolution reaction is:

$$Zn(OH)_2(s) + 2CH_3COOH(aq) \rightarrow Zn(CH_3COO)_2(aq) + 2H_2O(l) \qquad \text{(Eq. 9)}$$

Note that acetic acid is **not** written as $H^+(aq) + CH_3COO^-$ because it is a weak acid.

Reactions of the type shown in equation 9 occur because of the formation of H_2O by the reaction of hydrogen ion (H^+) from CH_3COOH and OH^- ion from $Zn(OH)_2$. This is shown in equation 10.

$$H^+(aq) + OH^-(aq) \rightarrow H_2O(l) \qquad \text{(Eq. 10)}$$

This reaction can only be written when the acid is strong, otherwise the net ionic equation will be as in equation 9.

Sometimes the reaction of an acid with an ionic solid produces additional compounds, other than just H_2O and a soluble compound. Consider the reaction of calcium carbonate ($CaCO_3$) with CH_3COOH solution:

$$CaCO_3(s) + 2CH_3COOH(aq) \rightarrow Ca(CH_3COO)_2(aq) + H_2O(l) + CO_2(g) \qquad \text{(Eq. 11)}$$

Solubility guidelines for compounds are established experimentally. We do this by observing the results of mixing specific solution combinations. We record our observations, including solution and precipitate color. Further, we need to test the solubility of all precipitates in CH_3COOH solution to establish the nature of the anion. We can then use our observations of the chemical behavior of known solutions to determine the identity of unknown solutions.

To illustrate the general procedure, consider aqueous solutions of five ionic compounds: sodium chloride ($NaCl$), lead(II) nitrate ($Pb(NO_3)_2$), sodium sulfate (Na_2SO_4), sodium carbonate (Na_2CO_3), and barium chloride ($BaCl_2$). These solutions are all clear and colorless, so they are not distinguishable by appearance. A systematic study the reactions of pairs of these solutions can be performed. We begin by putting three drops of $NaCl$ solution into each of four wells of a 24-well plate. We add three drops of $Pb(NO_3)_2$ solution to the first well, three drops of Na_2SO_4 solution to the second well, three drops of Na_2CO_3 solution to the third well, and three drops of $BaCl_2$ solution to the fourth well. While a white precipitate forms in the first well, the mixtures in the other three wells remain clear. We record this information as shown below, where we abbreviate white precipitate as "wt. ppt." and no reaction as "N.R."

Repeating this procedure with the other four solutions, we record the remaining data, as shown in Table 2. Examination of the data in Table 2 reveals that $NaCl$ solution only forms a white precipitate with one of the four other solutions. In contrast, lead (II) nitrate solution forms white precipitates with all of the other solutions. Both Na_2SO_4 and Na_2CO_3 form white precipitates with two of the other solutions, and $BaCl_2$ forms white precipitates with three of the other solutions. Given these data, if the labels were removed from our solutions, we could identify $NaCl$, $Pb(NO_3)_2$, and $BaCl_2$ based upon the number of white precipitates formed when the solutions are systematically mixed in pairs.

You should also note that we have mixed each pair of solutions twice: for example $NaCl + Pb(NO_3)_2$ and $Pb(NO_3)_2 + NaCl$. This is called a duplicate trial. The redundancy in the data is designed to minimize missing a result. If one trial produced a precipitate and the second one hadn't, we would repeat the results until we were sure which result was correct.

TABLE 2
RESULTS OF MIXING IONIC SOLUTIONS

Solution	NaCl	Pb(NO₃)₂	Na₂SO₄	Na₂CO₃	BaCl₂
$NaCl$	xx	wt. ppt.	N.R.	N.R.	N.R.
$Pb(NO_3)_2$	wt. ppt.	xx	wt. ppt.	wt. ppt.	wt. ppt.
Na_2SO_4	N.R.	wt. ppt.	xx	wt. ppt.	wt. ppt.
Na_2CO_3	N.R.	wt. ppt.	N.R.	xx	wt. ppt.
$BaCl_2$	N.R.	wt. ppt.	wt. ppt.	wt. ppt.	xx

We still cannot identify all the solutions. To distinguish between Na_2SO_4 and Na_2CO_3 solutions, we need more data. We acidify the precipitate-containing mixtures with CH_3COOH solution. The solubility or insolubility of a precipitate in CH_3COOH is indicated by the word "soluble" or "insoluble" in Table 3.

Now we see the usefulness of the CH_3COOH test. While neither of the precipitates formed by Na_2SO_4 is soluble in CH_3COOH solution, both of the precipitates formed by Na_2CO_3 are soluble in CH_3COOH. We can now distinguish the five solutions and write chemical equations describing the observed reactions. For example, the reaction of $Pb(NO_3)_2$ solution with Na_2CO_3 solution is shown in equation 12, and the reaction of the precipitate formed by the reaction with CH_3COOH solution is shown in Equation 13.

$$Pb(NO_3)_2(aq) + Na_2CO_3(aq) \rightarrow PbCO_3(s) + 2\ NaNO_3(aq) \tag{Eq.12}$$

$$PbCO_3(s) + 2\ CH_3COOH(aq) \rightarrow Pb(CH_3COO)_2(aq) + CO_2(g) + H_2O(l) \tag{Eq.13}$$

Now we repeat the procedure using the same solutions in unlabeled containers, labeled A, B, C, D, and E; and identify the solutions by systematically mixing them in pairs, again in duplicate. Our data might appear as shown in Table 4.

Based on a comparison of the data in Tables 3 and 4, we can identify our solutions. Solution A must be Na_2CO_3; it forms two acid-soluble precipitates with the other solutions. Solution B must be $BaCl_2$. It forms two acid-insoluble and one acid-soluble precipitate. Solution C is Na_2SO_4 because it forms two acid-insoluble precipitates. Solution D must be $Pb(NO_3)_2$ because it forms three acid-insoluble and one acid-soluble precipitate. Finally, Solution E must be NaCl because it forms one acid-insoluble precipitate; also because that's the only choice left.

In this experiment, you will work with six labeled solutions. First you will mix these solutions in pairs and test the solubility in CH_3COOH of any precipitates that form. Then you will work with the same

TABLE 3
SOLUBILITY IN CH₃COOH OF PRECIPITATES FORMED
AFTER MIXING PAIRS OF LABELED SOLUTIONS

Solution	NaCl	Pb(NO₃)₂	Na₂SO₄	Na₂CO₃	BaCl₂
NaCl	xx	wt. ppt. insoluble	N. R.	N. R.	N. R.
Pb(NO₃)₂	wt. ppt. insoluble	xx	wt. ppt. insoluble	wt. ppt. soluble	wt. ppt. insoluble
Na₂SO₄	N.R.	wt. ppt. insoluble	xx	N.R.	wt. ppt. insoluble
Na₂CO₃	N.R.	wt. ppt. soluble	N. R.	xx	wt. ppt. soluble
BaCl₂	N.R.	wt. ppt. insoluble	wt. ppt. insoluble	wt. ppt. soluble	xx

TABLE 4
SOLUBILITY IN CH₃COOH OF PRECIPITATES FORMED
AFTER MIXING PAIRS OF UNIDENTIFIED SOLUTIONS

Solution	A	B	C	D	E
A	xx	wt. ppt. soluble	N.R.	wt. ppt. soluble	N.R.
B	wt. ppt. soluble	xx	wt. ppt. insoluble	wt. ppt. insoluble	N. R.
C	N.R.	wt. ppt. insoluble	xx	wt. ppt. insoluble	N. R.
D	wt. ppt. soluble	wt. ppt. insoluble	wt. ppt. insoluble	xx	wt. ppt. insoluble
E	N.R.	N.R.	N, R.	wt. ppt. insoluble	xx

solutions in unlabeled containers. Your goal is to identify the contents of each container. Finally, based on your observations and the solubility guidelines, you will write chemical equations for the reactions you observe.

PROCEDURE

Chemical Alert: 1.0M acetic acid-irritant, 0.10M aluminum sulfate-irritant, 0.10M barium chloride-highly toxic and irritant, 0.10M calcium chloride-irritant, 0.10M magnesium sulfate-irritant, 0.10M sodium carbonate-irritant, 0.10M sodium hydroxide-toxic and corrosive, 0.10M sodium iodate-irritant and oxidant, 0.10M sodium oxalate-irritant, 0.10M sodium sulfate-irritant, 0.10M zinc nitrate-corrosive and oxidant

CAUTION: **Wear departmentally approved eye protection while doing this experiment.**

Handle all solutions with care. Do not ingest any of the solutions. Do not put foreign objects in reagent bottles. Do not return unused solutions to reagent bottles.

Note: You will be provided with 13 disposable Beral pipets: 6 for standard solutions; 6 for unknown solutions; and the last one for CH₃COOH solution.

1. Studying the Reactions of Known Solutions

Note: Your laboratory instructor will give you additional directions for labeling your well plate, test tubes, and Beral pipets.

1. Label six clean, dry test tubes with the name of the six solutions assigned to you. Label the seventh test tube "CH₃COOH solution." Place these test tubes in a test tube rack. Place 1–2 mL of the known solutions in the labeled test tubes.

2. Label six Beral pipets with the name of the six solutions assigned to you. Label the seventh pipet "CH₃COOH solution." Place these pipets in the corresponding test tubes. Be careful to avoid cross contamination. Small amounts of solution remaining in the pipets will be transferred to the test tube from which you draw the solutions. If the pipet is used in 2 different solutions, the second solution will be contaminated from the first solution. This will make any subsequent results suspect.

3. Using Table 2 as a model, enter the chemical identification of each of the six solutions in the table on Data Sheet 1.

Arrange your solutions in the test tube rack in front of you on the laboratory bench in the order in which you have designated them in the table on Data Sheet 1, with the CH₃COOH solution clearly separated from the six other solutions.

5. The columns and rows of your 24-well plate may be identified with letters and numbers, as shown in Figure 1. If they are not identified, label them as shown in the figure. You can then use the combination of a letter and a number to quickly identify any specific well. Thus, Well A-1 refers to the well in the first column (1) of the first row (A).

6. Using the appropriate pipet, place 3 drops of reagent in each well as designated.

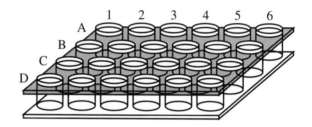

FIGURE 1. *Labeling a 24-well plate*

> reagent 1: all wells in column 1 and row A
> reagent 2: all wells in column 2 and row B
> reagent 3: all wells in column 3 and row C
> reagent 4: all wells in column 4 and row D
> reagent 5: all wells in column 5
> reagent 6: all wells in column 6

7. Grasp the base of the well plate and gently move it in small circles on the laboratory bench. Take care to prevent the solutions from swirling so fast that they splash out of the wells.

Note: Some reactions occur more slowly than others. Wait until you have observed a reaction mixture for 2–3 min before concluding that no reaction has occurred. Some precipitates are harder to see than others. Hold the plate over a dark surface and look directly down through the wells. If a contents looks hazy or cloudy, you have a precipitate. You may add a few more drops of either solution to attempt to produce a more obvious result. If you don't see a precipitate, you may want to add 3 drops of one reagent and 2 drops of another reagent. If it is too hot, you may not see a precipitate, and you may want to cool the test tube.

8. Carefully observe the mixtures. Record your observations in the table on Data Sheet 1. If a precipitate forms in any of the wells, describe its color.

9. Using the appropriate pipet, transfer a few drops of CH₃COOH solution to each well that contains a precipitate. Mix the solutions as you did in Step 7, and allow the mixtures to stand 2–3 min.

10. Observe any changes that occur in the precipitate after addition of the acid. Record your observations in the table on Data Sheet 1. If the precipitate dissolves, write "soluble." If the precipitate does not change, write "insoluble." If the precipitate appears to partially dissolve, wait 5 min, mixing the solution at 1 min intervals. If after 5 min the precipitate still remains partially dissolved, write "slightly soluble."

11. Dispose of the solutions in your well plate in the sink, using plenty of tap water; rinse twice with distilled water. It is not necessary to dry the well by wiping.

12. Repeat the procedure in Steps 6–11 using the following combinations of reagents.

    reagent 1: A1, B1
    reagent 2: A2, B2
    reagent 3: A3, B3
    reagent 4: A4, B4
    reagent 5: B5 and all wells in row A
    reagent 6: A6 and all wells in row B

    Record your observations on Data Sheet 1.

1	3	$BaCl_2$
2	1	$NaIO_3$
3	0	$NaCl$
4	2	$MgSO_4$
5		
6		

    Note: You have now tested each pair of solutions twice. Check your data to be certain that the results of both tests are in agreement for each combination. Repeat any test(s) for which your results are inconsistent. You may use test tubes or the well plate for repeat test. Just be careful not to contaminate your solution.

2. Studying the Reactions of Unknown Solutions

13. Obtain from your laboratory instructor a set of unknown solutions (prepared during the previous lab) in test tubes labeled 1–6. Following the procedure in Steps 2 and 3, label your pipets and fill in the table on Data Sheet 2.

14. Follow the procedure in Steps 6–12, using the unknown solutions.

    Record your observations for Solution 1 on your Data Sheet, for Solution 2, etc.

    Note: You have now tested each pair of solutions twice. Check your data to be certain that the results of both tests are in agreement for each combination. Repeat any test(s) for which your results are inconsistent. You may use test tubes or the well plate for repeat test. Just be careful not to contaminate your solution. Be careful not to use too much of your solutions. Your instructor can provide you with more, at a cost.

15. Based upon a comparison of the reaction patterns you observed for the known and unknown solutions, record on Data Sheet 3 the identity of each of the unknown solutions. Briefly cite the data you used as a basis for your identifications.

16. Discard the solutions remaining in your test tubes in the sink with plenty of tap water.

    Wash the test tubes with soap or detergent solution. Rinse each tube three times with tap water and once with distilled water. Dry the test tubes.

17. Dispose of all used pipets in the trash container.

    CAUTION: Wash your hands thoroughly with soap or detergent before leaving the laboratory.

$Pb(OH)_2$ $2Al(OH)_3 + CH_3COOH \rightarrow CO_2 + H_2O +$

$Al\ O_5\ H_7\ C_2\ AlO_2H + CH_4$ $C_1\ O_3\ H_2$ $Al\ CO_2H_5$

     $Al\ CH$ $C_1\ O_2\ H_5\ Al$

       $Al(CH_3COO)_3$ $KNaSO_4^{(s)} + KNO_3(aq)$

$CE - K_2SO_4(aq) + 2NaNO_3(aq) \rightarrow 2KNO_3(aq) + Na_2SO_4$

$NIE - K^+(aq) + SO_4^{2-}(aq) + Na^+(aq) \rightarrow KNaSO_4(s)$

$CE - CaBr_2(aq) + Na_2SO_4(aq) \rightarrow CaSO_4(s) + 2NaBr(aq)$

$NIE - \quad Ca^{2+}(aq) + SO_4^{2-}(aq) \rightarrow CaSO_4(s)$

$CE - Pb(IO_3)_2(aq) + 2NaOH(aq) \rightarrow Pb(OH)_2(s) + 2NaIO_3(aq)$

$NIE - \quad Pb^{2+}(aq) + 2OH^-(aq) \rightarrow Pb(OH)_2(s)$

$CE - K_2SO_4(aq) + BaCl_2(aq) \rightarrow BaSO_4(s) + 2KCl(aq)$

$NIE - \quad SO_4^{2-}(aq) + Ba^{2+}(aq) \rightarrow BaSO_4(s)$

$3Pb(ClO_4)_2 + Al_2(SO_4)_3 \rightarrow 3PbSO_4 + 2Al(ClO_4)_3$

$3Pb^{2+} + 6ClO_4^- + 2Al^{3+} + 3SO_4^{2-} \rightarrow 3PbSO_4 + 2Al^{3+} + 6ClO_4^-$

NaCrO4 $NaCl - Pb(ClO_4)_2$

$Pb(ClO_4)_2$ $Pb(ClO_4)_2 - NaOH$

$Al_2(SO_4)_3$ $\sim \ - Na_2CrO_4$

NaCl $\sim \ - Al_2(SO_4)_3$

           $NaOH - Al_2(SO_4)_3$

           $Na_2CrO_4 - Al_2(SO_4)_3$

NaOH

     $3Na_2CrO_4 + Al_2(SO_4)_3 \rightarrow Al_2(CrO_4)_3 + 3Na_2SO_4$

EMPIRICAL FORMULA OF COPPER CHLORIDE

BACKGROUND INFORMATION

Elemental analysis of a compound does not tell us how many of each atom exist in one molecule of that compound. Instead, we get the ratio of elements that occur. The simplest whole-number ratio in which kinds of atoms combine to form a compound is called the empirical formula of that compound. This ratio is fixed by the nature of the compound itself; sugar has a molar ratio of C:H:O of 1:2:1 whether it is chemically synthesized in a lab in Kansas or extracted from fruit in Australia.

Consider the reaction of 0.353 g of silvery-white magnesium metal (Mn) with an excess of atmospheric oxygen (O_2) to form magnesium oxide (Mn_xO_y). The unbalanced equation for this reaction is shown in equation 1, where x and y represent the number of Mn and of O atoms, respectively, that combine to form Mn_xO_y.

$$Mn(s) + O_2(g) \xrightarrow{\ heat\ } Mn_xO_y(s) \qquad \text{(Eq. 1)}$$

When we heat the carefully weighed sample of Mn in an open crucible weighing 17.208 g, a combustion reaction occurs. For this example,

$$\text{mass of } Mn_xO_y, g = 17.767\,g - 17.208\,g = 0.559\,g$$

We can find the empirical formula of this compound because we know exactly how much Mn it contains. Because Mn is the limiting reactant, the reaction continues until the supply of Mn is used. The amount of manganese used is given by equation 2.

$$mols\ X = \frac{mass\ of\ X}{molar\ mass\ of\ X} \qquad \text{(Eq. 2)}$$

$$mols\ Mn = \frac{mass\ of\ Mn\ used}{molar\ mass\ of\ Mn} = \left(\frac{0.353g}{54.94g/mol} \right) = 0.00643\ mol\ Mn$$

We determine the mass of O in the compound by subtracting the mass of Mn from the total mass of the compound, using equation 3.

$$\text{mass of O, g} = (\text{total mass of compound, g}) - (\text{mass of Mn, g}) \qquad \text{(Eq. 3)}$$

In our case, $m_O = 0.559g - 0.353g = 0.206g$. In general, one element's mass is found by subtraction. By applying equation 2, again, we find that the amount of O in the sample was 0.0129 mol. Because the empirical formula is the simplest whole number *ratio* of moles in a sample, we divide all the amounts by the smallest value. In our example, we have $0.00643/0.00642 = 1.00$ for Mn and $0.0129/0.00643 = 2$ for O. Thus, the ratio of Mn:O is 1:2, or the empirical formula is MnO_2.

In this experiment, you will determine the empirical formula of a compound of copper (Cu) and chlorine (Cl). You will add a known mass of zinc metal (Zn), a silvery-white solid, to a carefully measured amount of a blue solution containing a known amount of copper chloride. You will observe the reaction shown in equation 4.

$$Zn(s) + Cu_xCl_y \text{ (aq)} \rightarrow ZnCl_2(aq) + Cu(s) \qquad \text{(Eq. 4)}$$

As the reaction proceeds, the blue color will slowly disappear. As it does, you will see reddish-brown Cu metal forming in the solution. Once the reaction is complete and the solution is colorless, you will remove any unreacted Zn from the mixture and then separate the solid Cu from the solution. You will dry the Cu and determine its mass. From the original mass of copper chloride and the mass of Cu formed, you will determine the amounts of Cu and Cl in copper chloride. You will then calculate the empirical formula of the chloride.

From the empirical formula of copper chloride, you will be able to write an equation for the reaction of copper chloride solution with Zn. Based upon the stoichiometry of this equation and the amount of Zn used in the reaction, you will calculate the **theoretical yield** of Cu for the reaction. Finally, you will determine the **percent yield** for the reaction by comparing the experimental yield of Cu (**actual yield**) with the theoretical yield of Cu, using equation 5.

$$\% \ yield = \frac{actual \ yield}{theoretical \ yield} \times 100\% \qquad \text{(Eq. 5)}$$

Equipment

Aspiration Filtration: Using a Büchner Funnel

Büchner funnels are usually made of porcelain, but they can also be made of glass or plastic [see figure 1(a)]. A piece of filter paper or a glass fiber disk placed over the perforations in the funnel prevents the filtered precipitate from passing through the funnel. The funnel is placed over a side-arm flask. Suction supplied by a water aspirator or directly from a central vacuum (see figure 1(b)) system creates a partial vacuum that accelerates the filtration rate. Heavy-walled rubber tubing prevents tubing collapse during suction.

perforations

FIGURE 1a. *Büchner funnel.*

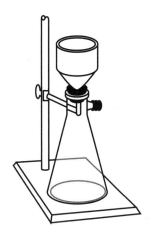

FIGURE 1b. *Suction filtration apparatus.*

Place a circular filter paper over the perforations in the funnel. The filter paper should be small enough to lie flat, yet large enough to completely cover the perforations. Moisten the paper with a small amount of distilled water from a wash bottle. Create suction by turning on the tap connected to the aspirator or turning on the vacuum line.

Once suction has begun, the vacuum filtration process is similar to gravity filtration: the bulk of the liquid is filtered first, then the bulk of the precipitate, and finally the rinse solutions. *Some liquid should be present in the Büchner funnel throughout the filtration.* After all the precipitate has been washed, continue to draw air through the precipitate to dry it somewhat, unless working with precipitates that react with atmospheric carbon dioxide. Do ***not*** turn off the water tap until filtration and washing are complete. Finally, remove hose from the side of the aspirator and ***then*** turn off the tap. Turning off the water first can draw the tap water into the flask through the pressure tubing.

PROCEDURE

Chemical Alert

copper chloride solution-toxic and irritant 10% hydrochloric acid solution-toxic and corrosive

CAUTION: **Wear departmentally approved eye protection while doing this experiment.**

CAUTION: Aqueous copper chloride is a toxic, irritating solution. Avoid contact with your eyes, skin, and clothing. Avoid ingesting the solution.

1. Obtain approximately 25.0 mL of copper chloride solution and transfer completely to a 150- or 250-mL beaker. Determine the mass of copper chloride in the solution from the concentration of the solution. Record the exact volume you used and the mass of copper chloride in the solution on your data sheet.

2. Obtain a flat piece of *clean* Zn and determine its mass to the nearest mg. Record this mass on your data sheet. The mass of Zn should be between 1.5 and 2 g. The zinc will probably have a piece of plastic backing. Carefully remove the backing. Try to avoid leaving fingerprints in the metal.

CAUTION: The reaction between Zn metal and copper chloride solution is quite exothermic. The beaker and the solution may become quite warm or even hot.

Note: In Step 3, avoid lowering the ends of the crucible tongs below the solution surface. Allow the Zn to slowly slide down the side of the tilted beaker.

3. Use crucible tongs to pick up the piece of Zn. Tilt the reaction beaker to a 45° angle. Release the Zn so that it slowly slides down the side of the beaker into the solution. Make sure that none of the solution splashes out of the beaker.

4. Continuously scrape the surface of the Zn piece in the reaction beaker with a clean, glass stirring rod, making sure that the solid Cu forming does not adhere to the Zn.

5. Allow the reaction to continue until the blue color has disappeared from the solution. Add 5 to 10 drops of 10% aqueous hydrochloric acid (HCl) to the solution, and thoroughly stir.

6. Remove the remaining Zn from the solution, using your crucible tongs. Transfer the Zn piece to a 50-mL beaker containing 5–7 mL of distilled water. Scrape off any adhering Cu and remove the cleaned Zn from the beaker. Repeat as necessary. Return all scraped off Cu, including water, to the beaker. Dry the remaining Zn with a paper towel or other absorbent paper. Determine the mass of the Zn remaining. Record this mass on your data sheet.

7. Discard this Zn in the container specified by your laboratory instructor.

8. Carefully decant the supernatant liquid (the aqueous solution in the beaker) into a 150-mL beaker from over the solid Cu, without losing any of the metal.

Note: In Step 9, do not allow any of the Cu in the reaction beaker to be lost.

9. Discard the supernatant liquid as specified by your laboratory instructor. Rinse the 150-mL beaker twice with 10 mL of distilled water. Discard this rinse water as instructed.

10. Add 10 mL of distilled water to the beaker. Transfer the water/Cu into a Büchner funnel installed onto a vacuum flask. Repeat the process until all visible traces of Cu are transferred from the beaker into the funnel.

11. Turn on the aspirator to start filtering. Wash the Cu with three 10-mL portions of acetone. Leave the aspirator running for 5 minutes to dry the Cu completely. Rinse the copper with several 5-mL aliquots of acetone to further dry it. **Caution: acetone is flammable. No flames may be used in the lab as long as anyone is drying samples with acetone.**

12. Transfer the Cu onto a piece of weighing paper and determine the weight of the paper and Cu. Record this data on your data sheet.

13. Leave the Cu on your bench top for another 5 minutes. Determine the weight of the weighing paper and Cu again. This should be within 0.05 g of the first determination (Step 12). If not, repeat this step until the last two readings agree within 0.05 g. Record the weight.

14. Discard the Cu residue in the container specified by your instructor.

15. Determine the weight of the weighing paper. Record this data on your data sheet.

Caution: Wash your hands thoroughly with soap or detergent before leaving the laboratory.

CALCULATIONS

Note: Letters refer to lines on data page (p. 50).

1. Determine the mass of Cu metal produced.   [e–d]

2. Determine the mass of Zn reacted.   [b–c]

3. Determine the amount of Cu in the copper chloride. [f/63.55]

4. Determine the mass of Cl in the copper chloride.   [a–f]

5. Determine the amount of Cl in the copper chloride. [i/35.45]

6. Calculate the empirical formula of the copper chloride.

7. Calculate the theoretical yield of copper from the amount of Zn used and the empirical formula.

8. Calculate the % yield from equation 5.

THE PHYSICAL DATA SEPARATION OF A TERNARY MIXTURE

INTRODUCTION

Chemists and industry are constantly separating m........ Sometimes it can be done physically and on occasion, chemical means are necessary. A *mixture* is matter containing two or more substances. Mixtures are classified as either *homogeneous* or *heterogeneous.* Pizza is an example of a heterogeneous mixture whereas air is an example of a homogeneous mixture. In a heterogeneous mixture it is possible to identify each part of the mixture. For example a pizza consists of cheese, sausage, mushrooms, green peppers, olives, etc. all of which retain their individual identities. In a homogeneous mixture it is not possible to identify each individual component. For example in air one cannot distinguish between the oxygen, the nitrogen, the water vapor, etc. When you separate your recyclables into glass, plastics, metals, paper, etc. you are physically separating a mixture. When you separate your laundry by colors, it is a physical process.

It is possible to separate a heterogeneous mixture because the items making up the mixture are not chemically reacting with each other. Each part of a heterogeneous mixture still has its fundamental properties and its own identity.

There are many methods for physically separating a mixture. Among the more prominent are:

a. Centrifugation   *This is the process of separating a solid or a precipitate from a liquid by whirling the mixture at a very high speed.*

In qualitative analysis, centrifugation is often used to separate a precipitate from the solution from which it was formed. When your washer is on the spin cycle, it is removing excess water from the clothes by spinning them (centrifugal force). When cheese is made, the whey is removed by centrifugal force.

b. Decantation   *This is the process of separating a liquid from a precipitate or a solid by carefully pouring off the liquid and leaving behind the solid.*

c. Distillation   *This is the process by which a liquid is brought to its boiling point, vaporized, then condensed and collected.*

If a mixture contains two or more liquids to be separated by distillation, they must have different boiling points. Because the boiling point of a liquid is normally determined at one atmosphere, all boiling points can be changed by elevating or by lowering the atmospheric pressure above the liquid. Distillation is very commonly used to purify water for home or laboratory use, to prepare spirits (vodka, port, bourbon, etc.).

Adapted from *General Chemistry Laboratory Manual.* Fourth Edition by D.L. Stevens. Copyright © 2004 by Kendall/Hunt Publishing Company. Reprinted by permission.

d. Extraction   *This is the process of separating one substance from another because it has a greater solubility in a different solvent.*

Vanillin is often extracted from the vanilla bean by placing the beans in ethanol. Because vanillin is more soluble in ethanol (ethyl alcohol) than water, the vanillin is "extracted" into the alcohol. When you purchase decaffeinated coffee, the caffeine has been extracted from the coffee beans using an organic solvent.

e. Filtration   *This is the process of removing a solid or a precipitate from a liquid by using a porous barrier, membrane, paper filter, etc. to prevent the movement of the solid or precipitate through the barrier.*

When you make coffee, the coffee filter prevents the solid from moving through the porous barrier.

f. Sublimation   *This is the process whereby a solid changes directly into a gas without a liquid phase being present.*

Dry ice (solid carbon dioxide, CO_2) sublimes. When dry ice is placed in the punch bowl at weddings, receptions, parties, etc., it chills the punch because during the physical change from a solid to gaseous state (no liquid state being present) energy is required (endothermic). The energy for this physical change comes from the liquid punch thus cooling it for the guests. The punch is not diluted because no liquids are created. The white vapors seen above the punch are condensing water vapor caused by the extremely cold temperature of the dry ice ($-78°C$). The vapors are not carbon dioxide gas as carbon dioxide is colorless. Some manufacturers package their food products in dry ice because of the extreme cold. Likewise, the food product will remain dry.

Iodine, I_2, also sublimes. If solid iodine is warmed, a purple gas will form immediately and then recrystallize on any nearby surface if in a semi-sealed system. Upon heating, ammonium chloride sublimes from a white solid into a white vapor.

Your mixture contains three components: a water soluble salt; a salt that dissolves in acid, but not water; and a completely insoluble material. You will separate the three components and identify the first two. To identify the ions that make up the salt, you will first perform a series of reference test so that you will know what a positive test for that ion looks like.

This experiment will take two weeks. Make sure you get the first separation done in the first week. Most students complete all the reference tests during the first week, and a few get as far as testing the first filtrate.

LABORATORY SAFETY

1. Departmental approved safety goggles and appropriate clothing are mandatory and must be worn while in the laboratory.

2. Because you will be using a Bunsen burner in this experiment, individuals with long hair should tie it back. Be very careful not to accidentally lean over a Bunsen burner while recording data. Keep your Bunsen burner back towards the center of the laboratory bench, in the hood or under the canopy. When not using your Bunsen burner, turn it off.

A FLOW CHART OF THE EXPERIMENTAL PROCEDURE

```
              ┌──────────────────┐
              │ Original Mixture │
              └────────┬─────────┘
                       │
              Add Hot water
         ┌─────────────┴─────────────┐
         ▼                           ▼
   ┌──────────┐                ┌──────────┐
   │ Filtrate │                │ Residue  │
   └────┬─────┘                └────┬─────┘
        │                           │
        ▼                       Add HCl
Test for anions and cations  ┌──────┴──────┐
                             ▼             ▼
                       ┌──────────┐  ┌──────────┐
                       │ Filtrate │  │ Residue  │
                       └────┬─────┘  └──────────┘
                            │
                            ▼
                     Test for cations
```

EXPERIMENTAL PROCEDURE

Part I: Separating the Water Soluble Component

You will be sharing the filtration apparatus. Work quickly. While one student separates, the other should begin the reference tests.

1. Weigh approximately 3 to 3.5 grams and record the exact mass to ±1 mg.

2. Boil* approximately 50 mL of distilled water. Pour about half of the water into the beaker with your sample. Stir thoroughly for several minutes.

   *If you have never used a Bunsen burner before see below.

HOW TO LIGHT A MICROBURNER

NEVER LIGHT A BURNER DIRECTLY UNDER GLASSWARE. MOVE IT TO ONE SIDE SO THAT THERE IS ONLY AIR ABOVE THE BURNER. Put the hose on the gas jet. Get a match out ready to strike. Turn the gas one half to three quarters of the way on. Light the match. Put the flame next to the barrel of the burner and slide it up until the gas ignites. Adjust the air-flow: the flame should be completely blue with a well defined inner cone. If there is yellow in the flame, open the air supply. If the flame is very loud ("roaring"), close the air down a little. When the flame looks good, adjust the gas flow so that the flame is the correct height. Adjust the gas slowly to avoid cutting the gas supply off completely.

When you have the flame completely adjusted, move it under the material to be heated.

3. Obtain a Büchner funnel and side-arm flask. Clamp the flask to a ringstand and put the funnel in the mouth of the flask. Connect the side arm to vacuum (either directly to a vacuum line, if one is available or to the side arm of a faucet for aspiration). If aspiration is to be used, turn the water on all the way (see figure below). Put a piece of filter paper in the funnel. Moisten it with distilled water. Make sure that the paper is covering all the little holes at the bottom of the funnel.

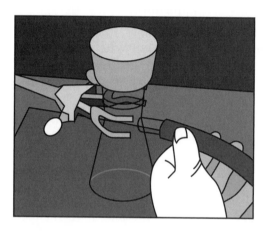 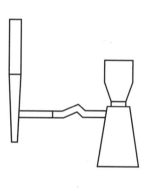

4. Carefully pour the mixture onto the filter paper. Use a stirring rod to direct the liquid to the center of the paper. Wash all the solid out of the beaker with distilled water. Pour the remaining hot water over the residue, dividing it in thirds. Leave the vacuum on for several minutes to draw as much liquid out as possible.

5. With the water still running (or the vacuum turned on) pull the hose off the side arm. If you turn the water off first, the flask still has a vacuum in it and it may pull tap water through the hose, contaminating your filtrate. Carefully remove the filter paper with the residue on it from the funnel and place it in a dry beaker. Set it aside to dry until next week.

6. Transfer the filtrate to another container (250-mL Erlenmeyer flask or Florence flask). You will test this solution for both cations and anions (see below).

Qualitative Tests for Ions

Following are listed a number of tests for common cations and anions. When testing your unknown, replace the unknown for the reference solution. For example, one test for Cu^{2+} requires that you add $NH_3(aq)$ to a $CuSO_4$ solution. If you replace the copper nitrate with your unknown and see the same behavior, you can conclude that the solution contains Cu^{2+}. All the concentrations listed below are approximate. If you can't find the same concentration, use what you have.

Cations

1. Na^+, K^+, and Ca^{2+}: Flame test. Some ions produce a distinctive color when placed in a flame. This is the reason that fireworks show the colors that they do. These colors depend only on the cations present. You will use this technique to identify Na^+, K^+, and Ca^{2+}. Before you begin, you must clean your nichrome wire. Light your burner. Place the loop of the wire in the blue flame about 0.5 cm over the top of the inner cone. This is the hottest part of the flame. The inner cone is actually the coldest (about room temperature). Wait several minutes until the flame is back to the blue color or else a dull orange. If the flame doesn't clean the wire sufficiently, dip the wire in a test tube filled with distilled water and swirl it around. Place the wire back in the flame and repeat two or three times or until clean.

If the wire is still not clean, put the wire in a test tube ⅓rd full of 2–3M HNO_3. Swirl and place in flame. Repeat as needed.

The flame test works best on solutions. You can do it with solids, but you will spend a lot of time cleaning your nichrome wire. Take 3 small test tubes, labeled Na, K and Ca. Place a small amount of $NaCl(s)$ in the first, $KCl(s)$ in the second and $CaCO_3(s)$ in the third. Cover the salts with enough dilute HCl or HNO_3 that you can see liquid over the salts. Swirl the test tubes *gently*. Heat the wire until it is glowing and quickly plunge it into the NaCl solution. Put it back in the flame. Note the color of the flame. Repeat with the other 2 solutions. If you are having trouble seeing K^+ and Ca^{2+}, hold a piece of cobalt glass to your eye while you put the wire in the flame. This will filter out the sodium color from the flame.

2. Ammonium

Reference solution: 0.1M NH_4Cl
Test reagent: 6M NaOH

Place 5 drops of the solution to be tested in a well of the microplate. Add 1–2 drops of the NaOH to the well. Hold a moistened piece of pH paper over the well. Be careful not to let the pH paper touch the well or any of the solutions. Check the pH of the gas over the well. Carefully smell the well. If the ammonium ion is present, the base will release it as ammonia gas. *no* change in pH paper
no precip

3. Cu^{2+}

Reference solution: 0.1M $CuSO_4$
Test reagents: 4M NH_3 or 0.1M $NaNO_2$

Many ions exist in solutions as complexes. Each Cu^{2+}, for example, is bound to 6 water molecules to form the ion $[Cu(H_2O)_6]^{2+}$. This complex ion has the distinctive blue-green color we expect from most copper solutions. Other ions or molecules can displace the water for the complex. Ammonia and nitrite form the complexes $[Cu(NH_3)_4]^{2+}$ and $[Cu(NO_2)_4]^{2-}$, respectively.

Place 3 drops of the solution to be tested in a microplate well. Add several drops of NH_3 and observe. Add a few more. Repeat with $NaNO_2$ in place of the NH_3.

For reference only, to give you a better understanding of what's going on, put 1–3 mL of $CuSO_4$ in a test tube. Add 4M NH_3 a few drops at a time. Watch very closely as you add the ammonia to the mixture both as the drops hit the test tube and as you mix the solutions together.

NH_3 - turned dark blue   *no precip*
$NaNO_3$ - none   *no precip*

4. Fe^{3+}

Reference solution: 0.1M $Fe(NO_3)_3$
Test reagents: 0.1M KSCN and 3M HCl

The iron (III) ion is light yellow in aqueous solution. Under acidic conditions, it forms a complex ion, $[Fe(SCN)]^{2+}$ when mixed with thiocyanate. Place 2 drops of the solution to be tested in a well. Add 1 drop of 3M HCl and 2 drops of KSCN and record your observations.

turned green-yellow
no precip

5. Ni^{2+}

Reference solution: 0.1M $NiCl_2$
Test reagents: 1% dimethylglyoxime in ethanol and NH_3

The Ni^{2+} ion forms a complex with dimethylglyoxime under basic conditions. Place 2 drops of the solution to be tested in a well. Add 1 drop of the dimethylglyoxime. If no reaction occurs, add 2 drops of NH_3. Record your observations.

turned dark-navy blue
no precip

6. Co^{2+}

Reference solution: 0.1M $Co(NO_3)_2$
Test reagents: NH_4SCN and acetone

Aqueous Co^{2+} exists as a complex ion $[Co(H_2O)_6]^{2+}$ which is pink. Thiocyanate in the presence of acetone can displace the water to form $[Co(SCN)_4]^{2-}$. Place 2 drops of the solution to be tested in a well. Add 5 drops of NH_4SCN and 5 drops of acetone. Record your observation. Make sure you add both reagents or you may confuse Co^{2+} with either Ni^{2+} or Cu^{2+}. Be careful with the acetone; it will damage the plastic the well plates are made of. Add the drops to the solution and avoid spilling it on the plastic.

turned yellow-gold color
turned darker w/ acetone
precipitated

Anions

NOTE: Study "Writing Chemical Equations" carefully. Pay special attention to the solubility rules. Remember: you do not need to test the second filtrate for anions.

1. NO_3^-

Reference solution: 0.1M KNO_3
Test reagent: Diphenylamine in H_2SO_4

Place 3 drops of the solution to be tested in a well. Add 3 drops of the diphenylamine solution to the well. Record your observations. **WARNING!** The diphenylamine is dissolved in concentrated sulfuric acid. It is very caustic. Avoid exposure to the skin. Handle with care!

2. SO_4^{2-}

Reference solution: 0.1M Na_2SO_4
Test reagent: 0.1M $BaCl_2$

Place 3 drops of the solution to be tested in a well. Add 1 drop of $BaCl_2$ to the well and record your observations.

3. Cl⁻, Br⁻, and I⁻

Reference solutions: 0.1M NaCl, 0.1M KBr, or 0.1M KI
Test reagents: 0.1M $AgNO_3$ and 4M NH_3

All halides react with Ag^+ to form precipitates. The salts formed can be distinguished on the basis of their color and solubility in aqueous ammonia. Ag^+ complexes with NH_3 to form $[Ag(NH_3)_2]^+$. Depending on the strength of the Ag-halide interaction, ammonia may not be able to dissolve the Ag^+. Place 3 drops of the solution to be tested in a well. Add 2 drops of water and 3 drops of $AgNO_3$. Record your observations. Add 5 drops of NH_3. Stir with a glass rod and record your observations. **NOTE:** when performing the reference test, comparison is easier if you set up the Cl⁻, Br⁻, and I⁻ at the same time in adjacent wells. When testing your unknown, only 1 well is needed.

4. CO_3^{2-} and S^{2-}

Reference solution: 0.1M Na_2CO_3
Test reagents: 6M HCl, Limewater (saturated $Ca(OH)_2$)

All carbonates and sulfides react with strong acids to form gases. Sulfides form H_2S which has the distinctive odor of rotten eggs. CO_2 is colorless and odorless. It does react with lime water (a saturated solution of $Ca(OH)_2$) to form $CaCO_3(s)$. Fill a large test tube about ⅓rd to ½ full with lime water. Try to avoid stirring up the sediment. In an ignition tube (looks like a test tube, but has thicker walls and no lip) place a few mL (fill it 2–3 cm deep) of the solution to be tested. Clamp the ignition tube at an angle as shown in the diagram. Add 4 drops of HCl to the tube. If the solution were a sulfide, even at a distance of several inches, you would smell the H_2S. Seal the end of the ignition tube with the bent glass tube. Put the other end of the tube in the limewater. Gently heat the HCl-solution mixture from the top down to drive the gas through the bent tube into the limewater. Observe and record your result.

Repeat the above tests with your solutions as needed to identify the ions in them.

PART II: Separating the Acid Soluble Component

Once again, you will be sharing the filtration apparatus. If you filtered first last week, your neighbor will go first this week. Work quickly and efficiently. **No flames will be permitted in the lab until EVERYONE has finished filtering!**

1. Carefully scrape the residue off the filter paper into a clean, dry weighed beaker. Record the mass on your data sheet.

2. Slowly add 10 mL of 3M HCl to your residue. Stir the solution with a clean stirring rod until all bubbling stops. Note any smell by gently wafting the gas towards your nose. Keep adding acid 1 mL at a time until no more bubbling occurs.

3. Stir in 1 mL more of acid to make sure all the acid soluble precipitate has dissolved. If the residue is colored, you may need more acid.

4. Set up the filtration apparatus as before and filter the mixture.

5. Wash the filtrate with two 2–3 mL aloquats of 3M HCl. Wash again with two 3 mL aloquats of distilled water.

6. Break the vacuum and transfer the filtrate (liquid) to another container. Re-establish a vacuum. Dry the residue by pouring two 5 mL aloquats of acetone through the residue. Keep the suction on for several minutes. **WARNING: ACETONE IS FLAMMABLE, EVEN MORE SO WHEN DISPERSED BY THE FILTRATION APPARATUS. IT IS IMPORTANT THAT NO FLAMES ARE IN USE IN THE LAB WHILE THE FILTRATION IS OCCURRING.**

7. Break the vacuum again and carefully transfer the residue to a beaker and weigh it. Give the filtration apparatus to your neighbor (if you are the first to filter).

Test the second filtrate for cations. Dispose of the residue, acetone and filtrates as instructed. Identify the components of your mixture and calculate the percent composition.

DETERMINING THE MOLAR VOLUME OF CARBON DIOXIDE

BACKGROUND INFORMATION

Ideal gases are gases in which the attractive and repulsive forces are negligible, and the volume of the gas particles is vanishingly small in comparison to the container. Avogadro's law states that equal volumes of ideal gases, under identical temperature and pressure conditions, contain equal numbers of molecules. Thus, at the same temperature and pressure, one mole of any ideal gas has the same volume as one mole of any other ideal gas. This volume is called the molar volume (V_m) of ideal gases. Under standard conditions (STP), which are a pressure of one atmosphere (atm) or 760 Torr, and a temperature of 273 kelvin (K), the molar volume of any ideal gas is 22.4 L.

The ideal gas law (Equation 1) expresses the relationship between pressure (P), volume (V), number of moles present (n), and temperature (T in kelvins) for any ideal gas sample.

$$PV = nRT \qquad \text{(Eq. 1)}$$

R is the gas law constant. So long as T, P, and V are expressed in kelvins, atmospheres, and liters, respectively, R equals 8.206×10^{-2} L atm mol^{-1} K^{-1} = 8.3145 J mol^{-1} K^{-1}.

If we measure the volume of a known mass of an ideal gas under one set of conditions from state 1 (labeled "1" below), and then change the conditions to state 2 (labeled "2"), we can use Equation 1 to write

$$n = \frac{P_1 V_1}{RT_1} \quad \text{and} \quad n = \frac{P_2 V_2}{RT_2}$$

Because the mass of gas does not change, n remains constant. Therefore, we can combine the two equations to form Equation 2 eliminating R on each side.

$$\frac{P_1 V_1}{T_1} = \frac{P_2 V_2}{T_2} \qquad \text{(Eq. 2)}$$

In order for Equation 2 to be valid, we must express the temperatures in kelvins. We can convert any temperature from degrees Celsius to kelvins by adding 273. In Equation 2, we can express the pressure in any of several units: inches or centimeters of mercury, Torr, or atmospheres. Similarly, we can express the volume in either milliliters or liters, as long as we use the same units for both states.

We can use Equation 2 and an ideal gas sample to determine the molar volume of that gas under standard conditions. For example, suppose a 0.250g sample of carbon dioxide (CO_2) gas has a volume of

0.145 L at 754 Torr and 27°C. To determine the molar volume of this sample at STP, we begin by solving Equation 2 for V_{STP}, where condition 2 is STP.

$$V_{STP} = \frac{P_1 V_1 (273K)}{T_1 P_{STP}}$$ (Eq. 3)

P_2 is listed as P_{STP}, rather than 1 atm, because you can use any unit of pressure you want. Next, we substitute the experimental values into Equation 3.

$$V_{STP} = \frac{(754)(0.145L)273K}{(27 + 273K)760Torr} = 0.131 \text{ L}$$

The resulting volume is not 22.4 L, as you might have expected, because the sample mass was only 0.250 g, much less than the molar mass of CO_2, 44.01 g/mol. However, we can use this result to calculate the molar volume of CO_2 at STP, using Equation 4.

$$\overline{V} = V_m = \frac{V_{STP}}{m_{sample}} \text{ (molar mass)}$$ (Eq. 4)

$$= \frac{0.131L}{0.250g \text{ } CO_2} (44.01g/mol \text{ } CO_2) = 23.1 \text{ L/mol}$$

We can find the percent error for this result relative to the theoretical ideal gas molar volume of 22.4 L/mol, using Equation 5.

$$\% \text{ error} = \frac{|Actual \text{ } value - Theoretical \text{ } value|}{Theoretical \text{ } value} \times 100\%$$ (Eq. 5)

$$\% \text{ error} = \frac{|23.1 - 22.4|}{22.4} \times 100\% = 3.13\%$$

In this experiment, you will calculate the molar volume of CO_2 gas (in L/mol) by determining the mass of a known volume of CO_2 gas at laboratory temperature and pressure. You will determine the CO_2 mass by first weighing a flask filled with air, and then reweighing the same flask filled with CO_2 gas.

You will fill the flask with CO_2 gas by allowing a piece of solid CO_2 (dry ice) to sublime in the flask. The $CO_2(g)$ will force all the air from the flask, because CO_2 gas is considerably more dense than air. The flask becomes filled with CO_2 gas, which we can assume is at laboratory temperature and pressure. The gas volume is the same in both cases; it is the volume of the flask.

The mass you will determine from the first weighing of the flask (m_1) is the mass of the flask (m_f) plus the mass of air contained in the flask (m_a), as shown in Equation 6.

$$m_1 = m_f + m_a$$ (Eq.6)

The mass you will determine from the second weighing of the flask (m_2) is the mass of the flask (m_f) plus the mass of CO_2 gas contained in the flask (m_{CO_2}) (Equation 7).

$$m_2 = m_f + m_{CO_2}$$ (Eq. 7)

We can calculate the difference between the second and first masses using Equation 8.

$$m_2 - m_1 = (m_f - m_{CO_2}) - (m_f - m_a) = m_{CO_2} - m_a \qquad \text{(Eq. 8)}$$

We can then calculate the mass of CO_2 gas using Equation 9, which is a rearrangement of Equation 8.

$$m_{CO_2} = m_2 - m_1 + m_a \qquad \text{(Eq. 9)}$$

As Equation 9 shows, the mass of CO_2 gas in the flask is equal to the difference between the two weighings plus the mass of air contained in the flask. We can calculate the mass of air, m_a, by multiplying the density of air at laboratory temperature and pressure by the volume of air contained in the flask (Equation 10).

$$m_a = d_{air} \times V_f \qquad \text{(Eq. 10)}$$

You can obtain the literature value for the density of air under your laboratory conditions by consulting a handbook of chemical data, such as the CRC *Handbook of Chemistry and Physics*. You will determine the volume of the flask by filling it with water. Based on your data, you will then calculate the molar volume of CO_2 gas. Finally, you will determine the percent error in your calculation, relative to the theoretical molar volume of CO_2 gas.

One important note: memorizing 22.4 L/mol has led more students astray than any other misconception about ideal gases. It is only valid at STP, and as you can see from this lab we are almost never operating under those conditions. It is a useful number for comparison purposes. It measures how close to ideality we are.

PROCEDURE

Chemical Alert: dry ice can cause frostbite.

CAUTION: Wear departmentally approved safety goggles while doing this experiment.

1. Carefully weigh an empty, stoppered 125-mL Erlenmeyer flask to the nearest 0.001g. Record the mass (m_1) of the flask, stopper, and contained air on your Data Sheet.

CAUTION: Dry ice can cause frostbite. Do not handle dry ice with bare hands. If necessary, use forceps to carry the dry ice to your work station.

2. Obtain a $\frac{1}{2}'' \times \frac{1}{2}'' \times \frac{1}{2}''$ piece (1.5–2.0 g) solid CO_2 (dry ice) from your laboratory instructor.

3. Place the solid CO_2 in the unstoppered flask.

4. When the solid CO_2 has completely vaporized, tightly stopper the flask.

5. Weigh the flask, stopper, and CO_2 gas. Record this mass (m_2) to the nearest 0.001g on your Data Sheet.

6. Do a second determination, recording all data on your Data Sheet. If time permits, do a third determination.

7. Unstopper the flask, and fill it to the brim with tap water.

8. Stopper the flask tightly, which will displace some of the water.

9. Using a towel or absorbent tissue, wipe any excess water from the outside of the flask.

10. Unstopper the flask. Carefully pour the water from the flask into a 100-mL graduated cylinder. Read the volume of water to the nearest milliliter and record, in liters, on your Data Sheet (3) as the volume (V_1) of the flask. The flask holds more than 100 mL so you will have to empty the cylinder and refill it with the remaining water to find the total volume of the flask.

11. Record the barometric pressure (P) on your Data Sheet. Be sure to indicate the units used.

12. Record the laboratory temperature (T_1), in °C, on your Data Sheet.

CAUTION: Wash your hands thoroughly with soap or detergent before leaving the laboratory.

CALCULATIONS

(Do the following calculations for each determination, and record the results on your Data Sheet.)

1. Convert your laboratory barometric pressure measurements to atmospheres, using Equations 11 and 12. Record the result on your Data Sheet.

$$1 \text{ atm} = 760 \text{ Torr} = 760 \text{ mm Hg} = 101.3 \text{ kPa} \qquad \text{(Eq. 11)}$$

$$2.54 \text{ cm Hg} = 1.0 \text{ in. Hg} \qquad \text{(Eq. 12)}$$

2. Convert the laboratory temperature in °C to kelvins. Record this temperature on your Data Sheet.

3. The density of air at STP is 1.29 g/L. Calculate the air density at the laboratory pressure and temperature, assuming an ideal gas behavior. Record the air density on your Data Sheet.

4. Determine the volume (in liters), V_2, of your CO_2 gas sample under standard conditions, using Equation 2 and your experimental data. Record this volume on your Data Sheet.

5. Calculate the mass of air contained in the flask (m_a), using Equation 10. Record this mass on your Data Sheet.

6. Calculate the mass of CO_2 gas (m_{CO_2}) contained in the flask, using Equation 9. Record this mass on your Data Sheet.

7. Find the number of moles of CO_2 gas (n) in your sample, using Equation 13.

$$n = \frac{m_{CO_2}}{MM_{CO_2}} \qquad \text{(Eq. 13)}$$

Record the number of moles of CO, on your Data Sheet.

8. Using Equation 14, calculate the molar volume $\overline{V}$ of CO_2 at STP and record it on your Data Sheet.

$$\overline{V} = \frac{V_2}{n} \qquad \text{(Eq. 14)}$$

9. Using the results of all of your determinations, calculate the average molar volume of CO_2. Record it on your Data Sheet.

10. Calculate the percent error for the average $\overline{V}$ as compared to the $\overline{V}$ of an ideal gas (22.4 L/mol), using Equation 5. Record this error on your Data Sheet.

92.9
47.9
140.8

LAB 7

REACTIVITY OF METALS

INTRODUCTION

In this experiment, the reactivity of a metal with hydrochloric acid will be determined. From its reactivity, the equivalent mass of the metal will also be determined. To obtain your experimental data, you will react a measured mass of metal with an excess of hydrochloric acid. You will collect the hydrogen gas produced by this reaction and note its temperature, volume, and pressure at collection time. This gas data will not be at *standard conditions*. With this data in hand, you will make the following calculations:

Determining Pressure of Hydrogen Produced

Using your experimental data, you will calculate the pressure of hydrogen gas. To do this, the pressure read off the barometer must be corrected. The hydrogen will be collected over water. In order to measure the volume, the hydrogen produced will displace water in a buret. The gas in the buret at the end of the experiment will be a mixture of hydrogen and water that evaporated into the hydrogen. In addition, the water level in the buret will be higher than the water level in the surrounding vessel. This column of water exerts a pressure called *hydrostatic pressure* and must be accounted for as well. Once the pressure of the hydrogen has been found, the number of moles of hydrogen can be calculated $PV = nRT$. This in turn tells us how many moles of H^+ were consumed reacting with the metal.

Determining Equivalent Mass of the Metal

The exact reaction will, of course depend on the metal involved. Reactive metals undergo a redox reaction, reducing H^+ to H_2 by being oxidized to M^{n+}. For example, the reaction might be:

$$M(s) + H^+(aq) \rightarrow \tfrac{1}{2} H_2(g) + M^+(aq)$$

or

$$M(s) + 2H^+(aq) \rightarrow H_2(g) + M^{2+}(aq)$$

or

$$2M(s) + 6H^+(aq) \rightarrow 3H_2(g) + 2M^{3+}(aq) \quad \text{etc.}$$

By measuring the hydrogen produced, we can only determine the number of moles of H^+ consumed. Without more information about the metal, we cannot know how many moles of metal were used. If we weighed the metal, however, we can determine what mass of metal will consume 1 mole of H^+. This mass is called the **equivalent mass** (sometimes called equivalent weight) because it tells us how much metal is equivalent to 1 mol of H^+. The equivalent mass is related to the molar mass by the equation below:

$$eq.\ wt. = \frac{molar\ mass}{n}, \text{ where } n \text{ is the number of electrons transferred in the process.}$$

LABORATORY SAFETY

In this experiment several safety precautions need to be observed. They include:

a. As always, approved safely goggles and appropriate clothes must be worn in the laboratory.

b. Since you will be working with hydrochloric acid (6 M) care must be exercised not to get this acid on your clothes or flesh. If you should come into physical contact with it, wash it off as fast as physically possible with copious amounts of water. Then inform your **teaching assistant of your spill.** *Speed is of the utmost importance in removing this acid from your clothes or flesh.* Do not tarry! **Avoid breathing the vapors of hydrochloric acid as they can irritate your lung tissue. Use the fumehood to keep the vapors of hydrochloric acid away from your lungs. On the raised part of the laboratory bench you may find a box of baking soda ($NaHCO_3$, sodium hydrogen carbonate or sodium bicarbonate). Use this white powder to neutralize any acid spills on the laboratory bench.**

c. **In Part D of the laboratory procedure, when you add the water to the acid in the buret, do so with caution as a violent reaction may occur due to the heat generated when the acid and water are mixed.**

d. **Handle your thermometer with care. When you remove it from your drawer be sure to hold both ends of the thermometer case so that it does not slip out.**

e. **Because of the inflammability of hydrogen, no flames of any type will be permitted in the laboratory while this experiment is being performed.**

EXPERIMENTAL PROCEDURE

Part 1—Calibrating the Buret

You will be using a buret as a eudiometer, an instrument used to measure gas volumes. Normally, a buret is used to dispense liquids through the stopcock. It is calibrated with 0.00 mL near the top and 50.00 mL near the bottom and can be read to the nearest 0.02mL. There is a stretch of glass between the 50.00 mL mark and the stopcock. We will be inverting the buret and allowing the gas to collect, starting at the stopcock and working its way down. We need to find the volume of the region between the stopcock and the 50.00 mL mark.

A. Obtain a buret and fill it with a few mL of water. The buret should read somewhere between 49 and 50. If the liquid is not over the 50 mL mark, add some more water to it. Record the initial volume on your data sheet.

49.1 mL

49.0 mL

B. Put an empty 10-mL graduated cylinder under the tip of the buret. Open the stopcock and let all the water drain into the graduated cylinder. Record the volume of water in the cylinder on your data sheet. You can now determine the uncalibrated or "dead volume" of your buret. There is a small error from the volume of water in the tip (which is *after* the stopcock), but this is small enough to safely ignore.

5.0 mL
4.9 mL
5.1 mL

Part 2—Measuring the Hydrogen Produced

A. Fill a 400/600 mL beaker three-fourths full of tap water.

B. Obtain a strip of metal ribbon that is about 2–2.5 cm long. If it has not been cleaned, clean it with a piece of steel wool. The reason it is necessary to clean the ribbon is because of the presence of the oxide, M_xO_y, and the nitride, M_aN_b, on the surface. Once cleaned, determine its mass. As your hands will leave body oils and deposits on the ribbon, you may wish to handle it with a pair of forceps once cleaned. Record this data. Do not use a piece with a mass greater than 0.0400 grams. If the piece of metal is heavier than 0.0400 g, cut and reweigh. Once massed, you may touch it without worry. Loosely roll the ribbon into a small ball. Place the roll in the center of a piece of plastic mesh. Fold the mesh around the spiral and tie it closed with a piece of copper wire. The resulting bundle should be just narrow enough to fit inside the buret. Bend the free end of the wire into a hook to catch the lip of the buret.

C. To your 50.00 mL eudiometer, carefully add about 5 mL of 6 M HCl. Keep the eudiometer at waist height while adding the acid so as to lessen the chance of spilling any acid on yourself. If at all possible, avoid breathing the HCl vapors.

5 mL

D. With the eudiometer tilted at about a 45 degree angle, add distilled water from a 250 mL or smaller beaker until the eudiometer is nearly full. Try not to mix the water and acid solution appreciably. Chemistry rules state that an acid should always be added to water and not visa versa. This is an exception to that rule.

E. At this point, insert your bundle into the eudiometer. Keep the tail of the wire outside of the eudiometer. Now completely fill the eudiometer to the brim with water. When full, the water should be beaded-up on the eudiometer's open end. Place the one-hole stopper into the open end, pinching the wire between the stopper and the wall of the eudiometer. The bundle should be several mm to 1 cm below the stopper, completely submerged in the water. Avoid leaving air in your eudiometer. With the eudiometer apparatus ready and taking care not to introduce any air, place a finger over the open end and carefully invert your eudiometer's open end into the beaker of water. Keep the open end of the eudiometer under the water's surface. If air enters into the eudiometer, you must begin again. That means you must start your experiment all over by cleaning and reweighing a new piece of metal, cleaning and refilling the eudiometer, etc.

F. Using a burette clamp, clamp the eudiometer in an upside down position with the open end of the eudiometer under the water (Figure 1). Leave some space between the beaker and the bottom of the eudiometer. Notice the movement of the more dense acid down the eudiometer towards the metal ribbon. Allow the reaction between the hydrochloric acid and metal to run its course.

Do not allow any air to enter the eudiometer. If the hydrogen gas generated exceeds the 50.00 mL volume mark on the eudiometer, it will be necessary to repeat the experiment but this time with a smaller mass of metal. Gently tap the eudiometer to get all the hydrogen to rise to the top and to verify that the reaction has gone to completion.

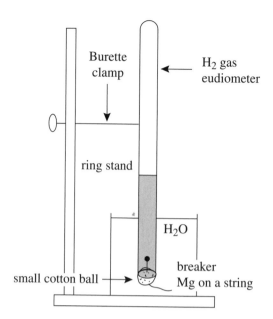

FIGURE 1

Once the reaction ceases (no visible evolution of hydrogen gas), carefully note the volume of the hydrogen gas collected. Tap the side so that all the gas rises to the top. Record this volume. Remember this gas was collected over water and is considered "wet" hydrogen.

G. Place a thermometer in the beaker of water to obtain the temperature of the collected hydrogen gas. We are assuming the temperature of the water and the temperature of the hydrogen gas are the same. Record this data.

H. With your ruler, measure the distance from the top of the water in the beaker to the bottom of the meniscus in the eudiometer. Record this data. Remove the eudiometer from the beaker. Take the bundle out, open it up to make sure all the metal reacted. If there is any metal left, weigh it and record this data.

I. On the wall is a barometer. Read the present atmospheric pressure from it. Record this data.

Since 1.00 mm of water is not equivalent to 1.00 mm of mercury, you must convert mm of water into mm of mercury. Pressure can be found from the mass of the column of liquid supported. In terms of a formula, $P = ghd$, where g is the acceleration due to gravity. We can convert from mm H_2O to Torr (mm Hg), by setting the pressures equal: $gh_1d_1 = gh_2d_2$. In this equation, h_1 will be the height of the column of water (in mm), d_1 is the density of water at this temperature, d_2 is the density of mercury (13.6 g/cm^3), and h_2 will be the pressure in Torr. The hydrostatic pressure must be subtracted from the atmospheric pressure.

J. Dispose of the waste as directed.

Dalton's Law of Partial Pressure states that the total pressure is the sum of the partial pressures. In this case the hydrogen gas was collected over water and thus contained water vapor. This hydrogen is said to be "wet". Thus the total pressure of the gas in the eudiometer was due to both the water vapor and the hydrogen gas at the temperature collected (plus the hydrostatic pressure). It can be expressed as follows:

$$P_{Total} = P_{Hydrogen\ Gas} + P_{Water\ Vapor} + P_{Hydrostatic}$$

As the vapor pressure of water is temperature dependent, you will be given the vapor pressure of water at your temperature. To correct the pressure of your hydrogen gas, the vapor pressure of water vapor *must be subtracted* from the original pressure. This will give you the pressure due solely to the "dry" hydrogen gas.

K. Using your experimental data, determine the number of moles of hydrogen produced. Then, using this result, calculate the number of moles of acid consumed. This, in turn is used to find the equivalent mass of the metal used. All of your calculations and set-ups must be shown and properly labeled to receive full credit.

TABLE 1
VAPOR PRESSURE OF WATER

Temp., °C	Vapor Pres. (Torr)	Temp., °C	Vapor Pres. (Torr)	Temp., °C	Vapor Pres. (Torr)
0	4.6	19	16.5	30	31.8
5	6.5	20	17.5	31	33.7
10	9.2	21	18.7	32	35.7
11	9.8	22	19.8	33	37.7
12	10.5	23	21.1	34	39.9
13	11.2	24	22.4	35	42.2
14	12.0	25	23.8	36	44.6
15	12.8	26	25.2	37	47.1
16	13.6	27	26.7	38	49.7
17	14.5	28	28.3	39	52.4
18	15.6	29	30.0	100	760

LAB 8

CHEMICAL KINETICS: METHYLENE GLYCOL CLOCK REACTION

PURPOSE OF THE EXPERIMENT

Study the kinetics of the reaction between methylene glycol and a bisulfite/sulfite solution.

Investigate how the rate of this reaction depends upon the initial concentrations of the reactants and upon the temperature at which the reaction takes place.

BACKGROUND INFORMATION

General

Thermodynamics asks *if* a reaction will occur and, if it does, to what extent it does. In answering this question, thermodynamics is concerned with the difference between the initial and final states of the reaction.

Kinetics, on the other hand, asks about the *rate* of a chemical reaction. It asks *how fast* the reactants can form products. Kinetics also is concerned with the *reaction mechanism*—that is, with the *route* the reaction takes between reactants and products. It explains how the system gets from its initial state to its final state.

Chemical reactions often occur in a sequence of steps, one of which is slower than any of the others. This slowest step is called the *rate-determining* or *rate-limiting step,* because it determines the overall reaction rate. Reaction products can't be formed any faster than the rate of this slowest step.

The rate of reaction can be expressed in terms of the concentrations of the reactants used. Such a mathematical formula is call a *rate law.* For the hypothetical reaction $aA + bB + cC \rightarrow dD + eE + fF$, the rate law is usually expressed in the form shown in equation 1. The symbol [ ] signifies the concentration of each species expressed in moles per liter.

$$rate = k[A]^x[B]^y[C]^z \qquad \text{Eq. 1}$$

The values of x, y, and z do not have to bare any relationship to the stoichiometric coefficients a, b, and c. Not all the reactant will have an effect on the rate, and so may not appear in the rate law. Therefore, the exponents x, y, and z must be determined experimentally. They are usually, but not always, small integers. They may be fractions, and they can even be negative. These exponents describe the order of the reaction. For example, in equation 1, if $x = 1$, $y = 3$, and $z = 0$, the reaction is first order with respect to A, third order with respect to B, and C does not appear in the rate law. The overall order of the reaction is fourth, the sum of the exponents of all the compounds in the rate law.

The specific rate constant, k, depends primarily on the nature of the reaction in question. In general, k is independent of the concentration of the reactants. However, under extreme conditions, k may vary at very high or very low concentrations. This usually reflects a change in the reaction mechanism. The rate constant is only valid at a given temperature, because it has a temperature dependence.

Any step in a sequence of steps making up a reaction can be illustrated on a diagram like the one in Figure 1. In order for this single step to occur, the reactants must go over the energy barrier separating them from products. It is *the ability of the reactants to get over this barrier in the rate-determining step that determines the rate of the reaction.* The height of the barrier is called the activation energy—the energy required to get this reaction step started (see figure 1). As the temperature is increased, the number of collisions with enough energy to get over the barrier increases. Thus, we expect reaction rates to increase with increasing temperature. This can be summarized by the Arrhenius equation (equation 2), named after the Swedish chemist Svante Arrhenius (1859–1927) who first described this behavior.

$$k = Ae^{-\frac{E_a}{RT}}$$

(Eq. 2)

In equation 2, k is the specific rate constant, T is the temperature in kelvin, A is the collision frequency factor for the reaction, and R is the ideal gas constant. The A factor, which has the same units as k does, depends on the frequency with which the reactants collide in the right orientation or geometry. For the reaction to occur, the molecules must "bump into" each other in the "right way," forming what is called an *activated complex.* Not only do they have to have the correct energy, but they must also have the right orientation.

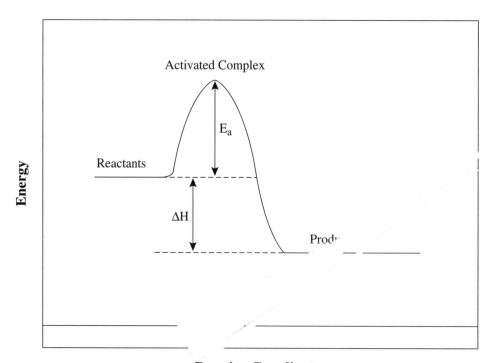

Reaction Coordinate

FIGURE 1

The Chemical Reaction to Be Studied

A *clock reaction* is a reaction in which some visible change takes place in the reaction mixture when a certain point in the reaction is reached. For example, an acid-base indicator like phenolphthalein can be used to signal a point at which the pH of the solution reaches a certain value. Clock reactions are useful in studying reaction kinetics because time is inversely proportional to rate. The longer a reaction takes the slower the rate must be and vice versa.

The reaction you will study in this experiment is called the formaldehyde clock reaction, but is more accurately a methylene glycol clock. The overall reaction is between methylene glycol ($CH_2(OH)_2$) and sodium bisulfate in water. The mechanism for this reaction is actually quite complicated, but can be simplified as given below.

Step 1: $CH_2(OH)_2 \leftrightharpoons HCHO + H_2O$ Slowest

Step 2: $HCHO + SO_3^{2-} \rightarrow H_2CO(SO_3)^{2-}$ Slow

Step 3: $H_2CO(SO_3)^{2-} + H^+ \rightarrow H_2COH(SO_3)^-$ Fast

Step 4: $HSO_3^- \leftrightharpoons H^+ + SO_3^{2-}$ Fast

Overall: $CH_3(OH)_2 + HSO_3^- \rightarrow H_2COH(SO_3)$ Eq. 3

This clock reaction has been described by M. G. Burnett *J. Chem. Educ.*, 59, 160 (1982). The rate determining step is the formation of formaldehyde in step 1. Once the formaldehyde is generated, it will react with the sulfite ion (SO_3^{2-}) to form an adduct ($H_2CO(SO_3)^{2-}$). Subsequent addition of hydronium ion (step 3) is faster than step 2. Since the HSO_3^-/SO_3^{2-} mixture (weak acid/conjugate base) forms a buffer solution, the hydronium ion is quickly replenished through the rapid equilibrium in step 4. This buffering effect will disappear when the HSO_3^- is almost completely depleted by the reaction (equation 3). Phenophthalein indicator in the solution detects this sudden change in pH resulting in the reaction mixture suddenly becoming basic and the color of the solution changes rapidly from clear to pink.

The initial rate of the reaction depends upon both the bisulfite concentration and the methylene glycol concentration. A change in either one may affect the reaction rate.

Determining the Rate Equation

The rate of a chemical reaction can be followed by measuring the appearance of product, or the disappearance of reactants as a function of time. Since the reaction rate is a function of the reactant concentrations, the rate of a given reaction is given by equation 4:

$$Rate = \frac{d[product]}{dt} = -\frac{d[reactant]}{dt} \qquad (Eq. 4)$$

The rate equation can be learned only by experiment.

In this experiment you will use the **method of initial rates** to find the rate law and the reaction orders for methylene glycol-bisulfite reaction. As a reaction proceeds, reactants are used up and therefore their concentrations decrease. Consequently, the rate of reaction decreases. The method of initial rates assumes that the reaction rate is measured before the reaction has slowed to any significant extent.

Throughout this experiment, the concentration of methylene glycol is kept in excess by 30 to 100 times. Thus, the concentration of methylene glycol does not change significantly even if the HSO_3^- is completely consumed by the reaction. Also, the concentration of HSO_3^- is kept nearly constant until the end of the reaction through the buffering effect. The assumptions of the initial rate method are valid, since the concentrations of reactants will remain relatively constant until the visible change (clock) appears.

With this additional information, the reaction rate can be written in the form of equation 5.

$$rate = \frac{[HSO_3^-]_0}{time\ needed\ for\ color\ change} \qquad \text{(Eq. 5)}$$

The value $[HSO_3^-]_0$ is the initial concentration of bisulfite.

The orders of the reaction (exponents a and b) reflect the dependence of the reaction rate on the concentrations of the two reactants, shown in equation 6.

$$\text{Rate} = k[\text{methylene glycol}]^a\ [\text{bisulfite}]^b \qquad \text{(Eq. 6)}$$

If you take the logarithm of both sides of the rate equation, then you get equation 7.

$$\log(\text{rate}) = \log(k) + a\log[\text{methylene glycol}] + b\log[\text{bisulfite}] \qquad \text{(Eq. 7)}$$

If the concentration of one of the reactants (e.g., bisulfite) is kept constant, then the term $b\log[\text{bisulfite}]$ becomes a constant and a plot of log(rate) against log(methylene glycol) should produce a straight line with a slope of a. Likewise, if the concentration of the other reactant (methylene glycol) is kept constant, then a plot of log(rate) against log(bisulfite) should produce a straight line with a slope of b.

Calculating the Activation Energy and Frequency Factor

In this experiment you will also study the effect of temperature upon the reaction rate. The Arrhenius equation (equation 2) describes the dependence of the rate constant k on temperature. Taking the natural log of both sides of equation 2 turns the Arrhenius equation into equation 8.

$$\ln k = \ln A - \left(\frac{E_a}{R}\right)\left(\frac{1}{T}\right) \qquad \text{(Eq. 8)}$$

E_A is the activation energy, R is the gas constant, T is the temperature (K), and A is the frequency factor. This equation illustrates the three factors that govern the rate constant:

1. The **temperature, T:** the higher the temperature, the faster the rate.

2. The **activation energy, E_a:** the higher the activation energy, the slower the rate.

3. The **collision frequency factor, A:** reflects the ease of formation of the activated complex.

If the rate constants are measured at different temperatures while keeping the initial concentrations constant, the temperature dependence of the rate constant is readily apparent. A plot of $\ln k$ against $1/T$ gives a straight line with a slope of $-E_a/R$ and a y-intercept of $\ln A$.

PROCEDURE

This experiment should be done with a partner.

Warning! Wear proper eye protection throughout this experiment.

Warning! Do not use tap water for this experiment. If the water is impure, your reaction may not proceed correctly.

Caution! Formaldehyde is an irritant of both skin and lungs. Clean up all spills carefully and work in the hood as much as possible.

Part I: Determination of the Order of the Reaction

1. Fill a 600-mL beaker with distilled water. To ensure a fairly constant temperature, use only this water for the first group of experiments.

2. Prepare solution A by dissolving 4.0 g of sodium bisulfite and 0.6 g of sodium sulfite in 250 mL of distilled water. Weigh these substances carefully to the nearest milligram on an electronic balance and record the masses on the data sheet. Solution B, the methylene glycol solution, has been prepared for you. Throughout your experiment, you must take samples from the same solutions of A and B.

3. Collect 200 mL of solution B in a 400-mL beaker. Make sure you label the beaker first. This should be sufficient solution for you and your partner to complete part 1. **Penalties may be assessed if you require additional solution.**

4. Label one 10-mL graduated cylinder "A" and another "B." You will use these to transfer the solutions to the reaction flask.

5. Fill a 100-ml graduated cylinder as precisely as possible with the volume of distilled water shown in Table 1. Add this water to a 250-mL Erlenmeyer flask.

6. Add the appropriate volume of solution A to the Erlenmeyer flask; add 10 drops of phenolphthalein indicator and swirl to mix the solution. Add the appropriate amount of solution B to the cylinder labeled "B" (use a larger cylinder for trials 7 and 8).

7. In this step, one partner is responsible for timing the reaction to the nearest second while the other partner is responsible for mixing the solutions. With one partner checking the time, the other partner pours the solution from cylinder B prepared in step 6 into the reaction flask and swirl to mix the solutions thoroughly. Record on the data sheet how many seconds it takes for the pink color to appear. If the color change takes you by surprise, repeat the trial. It is important to start timing from the moment of mixing and not necessarily from the time your partner says "now!"

8. Rinse the flasks thoroughly with distilled water and shake out any excess water. Do all trials 1 through 8. Enter all the data on the data sheet.

9. Repeat trials 1 though 8, take the averages of the times recorded, and then calculate the ratio of [HSO_3^-] to the average time to obtain a measure of the reaction rates. Enter all these data on the data sheet.

Dispose of all waste properly as instructed.

TABLE 1
COMPOSITIONS OF SOLUTIONS

Trial	Distilled Water	A (mL)	B (mL)	Total Volume (mL)
1	90.0	5.0	5.0	100
2	85.0	5.0	10.0	100
3	80.0	5.0	15.0	100
4	75.0	5.0	20.0	100
5	87.5	2.5	10.0	100
6	80.0	10.0	10.0	100
7	75.0	15.0	10.0	100
8	70.0	20.0	10.0	100

Part II: Determination of the Activation Energy and the Frequency Factor

1. Collect an additional 100 mL of solution B.

2. Fill a 600-mL beaker with approximately 200 mL of room temperature water.

3. Prepare the solutions according to trial 1 from table 1 (see steps 5 and 6 from part I). This time, keep the Erlenmeyer flask immersed in the water in the beaker.

4. Measure the reaction time as in part I step 7. Be careful in mixing not to splash the water into the flask. Record the time on the data sheet. Measure the temperature of the reaction mixture and record it on the data sheet. Repeat again, record the time, the average time and the rate as in step 8 of part I.

5. Replace the water in the 600-mL beaker with an ice-water mixture and repeat steps 3 and 4, above.

6. Using an electric heater plate, heat water to 35–40°C and repeat steps 3 and 4.

7. Repeat again at a higher temperature.

Calculations

The method of initial rates allows you quickly to spot the approximate relationship between the rate and the concentration of each reactant. If, for example, one reactant concentration is held constant and the concentration of the other reactant is doubled, and if it is observed that the rate also doubles, then the rate must be directly proportional to the concentration of that reactant. In other words, the order with respect to that reactant must be 1. In another example, if the concentration of one reactant is doubled and the rate is observed to increase by a factor of four, then the order with respect to that reactant must be two (since $2^2 = 4$). From your data you should be able to estimate the orders with respect to both [A] and [B]. Enter these estimates on your data sheet.

Calculate the concentration of methylene glycol and bisulfite in each trial. The concentration of the methylene glycol will be given to you. Calculate the concentration of $NaHSO_3$ in the stock solution from equation 9.

$$[NaHSO_3] = \frac{\dfrac{mass\ NaHSO_3}{MW}}{0.250L} \qquad\qquad Eq.\ 9$$

The concentration in each trial merely accounts for the dilution factor. Because the final volume is 100 mL in all cases, the concentration of bisulfite or of methylene glycol is given by equation 10

$$\frac{volume\ added\ (mL) \times concentration\ (M)}{100\ mL} \qquad\qquad Eq.\ 10$$

Draw a graph of the log (rate) against log [A] for trials 5 through 8. On another piece of graph paper, draw a graph of log (rate) against log[B] for trials 1 through 4. Draw the best straight line through points 1 through 4 and another line through points 5 through 8. Measure the slope of each line and record the results on the data sheet. Determine the orders of reaction; then calculate the overall order. Enter the results of these calculations on your data sheet.

Pick any concentration and find the specific rate constant. The rate law is given by equation 4. So, the rate constant can be found from equation 11. Enter this value on your data sheet.

$$\frac{rate}{[A]^a[B]^b} \qquad\qquad Eq.\ 11$$

For the temperature dependence, calculate the specific rate constant k for each temperature and take its natural log (ln). Convert the temperature to kelvin and calculate $1/T$. Plot ln k (y-axis) against $1/T$ (x-axis) for each temperature you measured. Find the best-fit straight line through these points. The slope of that line is $-E_a/R$, and the y-intercept is ln(A). To express the activation energy in SI units, we use 8.314 J mol^{-1} K^{-1} for R.

$$rate = k[A]^x[B]^y[C]^z$$

$$\frac{2.7 \times 10^{-6}}{1.1 \times 10^{-5}} = \frac{k[0.010]^x[0.25]^y[0.10]^z}{k[0.020]^x[0.25]^y[0.10]^z}$$

$$0.2454 = \frac{1}{2^x}$$

$$2^x = 4.07$$

$$x = 2$$

$$3^x = 8.14$$

$$\frac{2.2 \times 10^{-5}}{2.7 \times 10^{-6}} = \frac{[0.030]^x}{[0.010]^x}$$

$$\frac{2.2 \times 10^{-5}}{1.1 \times 10^{-5}} = \frac{[0.030]^x}{[0.020]^x}$$

$$2 = 1.5^x$$

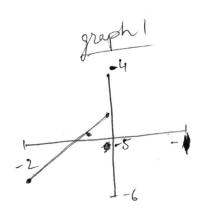

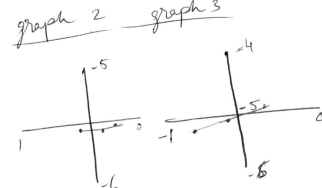

graph 1   graph 2   graph 3

LAB 9
ENTHALPY OF FORMATION OF AMMONIUM SALTS

BACKGROUND INFORMATION

There are a very large number of chemical reactions that have never been investigated. There may be good reasons for not performing certain chemical reactions. For example, some reactions may pose great risks to the environment or to living organisms. Other reactions may be too costly in terms of time or money. Studying the chemical reactivity of diamonds would certainly be an expensive undertaking. Fortunately, there are methods that allow us to study certain aspects of chemical reactions without ever having to perform the reactions.

Hess's Law (also known as the *Law of Constant Heat Summation*) is an example of a method that allows chemists to study the **thermodynamics** of a chemical reaction without having to perform the reaction itself. Germain Henri Hess discovered this fundamental law of nature in 1840 while studying the heat evolved when acids and bases are mixed with water. Hess's Law states that the **enthalpy change (ΔH)** of individual steps in a process can be added or subtracted to determine the overall enthalpy change of the process. Hess's Law is a direct consequence of the **First Law of Thermodynamics,** which requires that the enthalpy change of a reaction be the same regardless of the reaction pathway that leads from reactants to products. Thus, the net amount of heat liberated or absorbed in a chemical process is the same, regardless of whether the process is performed in one step or multiple steps.

Because of the additive property of ΔH, we can use a standard set of reactions from which we can generate any other reaction. All we have to do is add the enthalpies of these standard reactions and we have the enthalpy of our reaction of interest. The standard reactions we use are called **formation** reactions and the enthalpies are **enthalpies of** formation symbolized by ΔH_f^o. A formation reaction has one single product: one mole of the compound. The reactants are elements in their standard states. Thus, on paper, we can decompose any compound to its elements ($\Delta H = -\Delta H_f^o$ and then take those elements and rearrange them into the product compounds ($\Delta H = \Delta H_f^o$). This leads to the equation 1.

$$\Delta H^o = \sum_{Product} \Delta H_f^o - \sum_{Reactants} \Delta H_f^o \tag{1}$$

This process for the reaction $3C_2H_2 \rightarrow C_6H_6$ is shown in figure 1.

In today's experiment, you will use Hess's Law to determine the enthalpy change (ΔH) of a chemical reaction that cannot be easily measured by experimental methods. This reaction is represented by equation (2).

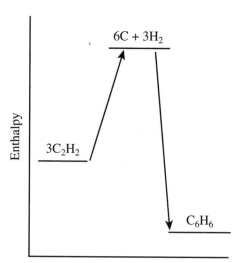

FIGURE 1. *Enthalpy of conversion of acetylene to benzene.*

$$\frac{1}{2}N_2(g) + 2H_2(g) + \frac{1}{2}Cl_2(g) \rightarrow NH_4Cl(s) \text{ or } N_2(g) + 2H_2(g) + \frac{3}{2}O_2(g) \rightarrow NH_4NO_3(s) \quad \Delta H_1 \quad (2)$$

You will determine the value of ΔH for this reaction by measuring the enthalpy changes of two related chemical reactions. These reactions are represented by equations (3) and (4) below.

$$NH_3(aq) + HCl(aq) \rightarrow NH_4Cl(aq) \text{ or } NH_3(aq) + HNO_3(aq) \rightarrow NH_4NO_3(aq) \quad \Delta H_2 \quad (3)$$

$$NH_4Cl(aq) \xrightarrow{-H_2O} NH_4Cl(s) \text{ or } NH_4NO_3(aq) \xrightarrow{-H_2O} NH_4NO_3(s) \quad \Delta H_3 \quad (4)$$

Equation (2) can be obtained by adding equation (3) to equation (4) along with the enthalpies of formation of the aqueous and ammonia (shown in equations (5) and (6), below).

$$H_2(g) + \frac{1}{2}Cl_2(g) \xrightarrow{H_2O} HCl(aq) \; H_2(g) + \frac{1}{2}N_2(g) + \frac{3}{2}O_2(g) \xrightarrow{H_2O} HNO_3(aq) \quad \Delta H_3 \quad (5)$$

$$\frac{1}{2}N_2(g) + \frac{3}{2}H_2(g) \xrightarrow{H_2O} NH_3(aq) \quad \Delta H_4 \quad (6)$$

We simply add the enthalpies for (3), (5) and (6) together with the reverse of (4)

$$\Delta H_f^o = \Delta H_3 + \Delta H_4 + \Delta H_5 + \Delta H_6 \tag{7}$$

The enthalpies for (5) and (6) are readily available from the literature.

Equation (1) can be obtained by adding equation (2) to the reverse of equation (3). If we reverse equation (3), we must also change the sign of the enthalpy change for the reaction. According to Hess's Law, if we add the equations for two reactions together, we can also add their enthalpy changes together. Thus, the enthalpy change for reaction (1) is simply the sum of the enthalpy change for reaction (2) and the reverse of reaction (3).

$$\Delta H_1 = \Delta H_2 + (-\Delta H_3) \tag{8}$$

To determine the enthalpy changes (ΔH_2 and ΔH_3) of reactions (2) and (3), you must measure the **heat (q)** released by each reaction. You will do this using the same techniques and concepts you used in the experiment on calorimetry. (You might want to review the Introduction to *Calorimetry.*) Recall that the heat exchanged in a process can be calculated from the following equation

$$q = c \times m \times \Delta T \tag{9}$$

where m is the mass of the substance in grams, the **specific heat,** *c,* of the substance is expressed in units of J/g•°C, and ΔT is the temperature change the substance undergoes ($T_{final} - T_{initial}$). Some heat will be lost through the walls and lid of the calorimeter. A correction term $C_c = 15$ J/K will be used to estimate this loss. Our corrected relationship between heat and temperature becomes: $q = (c \times m + 15)\Delta T$. The specific heat capacities can be found in Table 2.

$$q_{water} = (4.18 \text{ J/g•°C})(m_{water})(\Delta T_{water}) \tag{10}$$

The heat gained by the water is equal in magnitude by opposite in sign to the heat lost by the reaction. Therefore, we can calculate $q_{reaction}$ from the following relationship.

$$q_{reaction} = -q_{water} \tag{11}$$

Finally, the enthalpy change of each reaction, $\Delta H_{reaction}$, can be determined by dividing the heat lost by the reaction ($q_{reaction}$) by the number of moles of product formed (either NH_4Cl(aq) or NH_4NO_3(aq)). For reaction (2), this calculation takes the following form.

$$\Delta H_3 = \frac{q_3}{moles\ NH_4Cl\ or\ NH_4NO_3} \tag{12}$$

PROCEDURE

CAUTION: Wear departmentally approved eye protection. *If you are found without the eye protection, penalty will be imposed without warning.*

This experiment will be performed in pairs. The write-up, calculations, graphing, etc., must be done totally independently. A series of two experiments will be performed. Assume the calorimeter constant, C_C, to be 15J K^{-1}.

Part A.   Determination of the enthalpy of reaction of aqueous ammonia with a strong acid (HCl or HNO_3) assigned by the instructor.

Part B.   Determination of the enthalpy of solution of the ammonium salt of the acid used in Part A (NH_4Cl or NH_4NO_3).

In addition to reagents, the equipment needed is: two 6-oz pressed Polystyrene cups (and one cup lid), a 150-mL beaker, two 100-mL graduated cylinders, and one thermometer. Keep the cups together, as though they were glued together. You need 2 cups for the additional insulation.

Solution temperature as a function of time will be recorded on the data sheet. In addition to recording the data, a temperature-time plot on graph paper should be prepared; this can be done while the data are being collected. The temperature should be plotted on the y-axis and the time in minutes on the x-axis. The T observed will be between 7 and 14° for Part A and between –1 and –6° for Part B. The scale units on the y-axis should be selected so that all points can be plotted on scale; the scale units on the x-axis should cover the range 0–20 minutes.

The procedure is essentially the same for both parts:

1. Rinse one Polystyrene cup and beaker with distilled water and wipe dry with a clean piece of absorbent paper. Nest it in the other cup (see figure 2). Clean a beaker with distilled water and wipe dry with absorbent paper. Place the thermometer through the lid of the cup with the rubber ring over the hole set so that the thermometer will not touch the bottom of the cup when the lid is placed on the cup. Do not clamp the thermometer to the ring stand. They are fragile and will likely break if clamped.

2. Using graduated cylinders, measure solution 1 (see Table 1) into the cup and (except in Part B) measure solution 2 into the beaker. Record the concentrations of the solutions used on the Data Sheet.

Note: Calculate the concentration of the ammonium salt—NH_4Cl or NH_4NO_3—formed in the reaction of Part A. Determine the mass of salt in 100 mL of solution; weigh out that amount of salt and use in place of solution 2 in Part B.

3. Check the temperature of solution 1. Assume the temperature of the two solutions before mixing is essentially the same. Record the temperature of solution 1 at 1-minute intervals for 5 minutes.

4. After 5 minutes of temperature readings, pour solution 2 into the calorimeter cup. Immediately swirl the cup gently to mix the solution well. Record the time of mixing. (Never use a thermometer as a stirring rod!) Continue to record the temperature at 1-minute intervals for 10 minutes.

Do one determination of the temperature change for each part of the experiment.

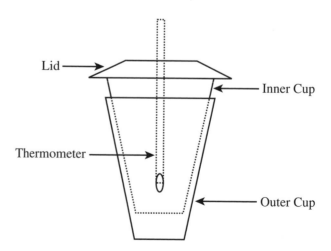

FIGURE 2. *Nested polystyrene cup.*

TABLE 1
REAGENTS TO BE USED

Part	A	B
Solution 1	50.0 mL acid (2.00M HCl or 2.00M HNO₃)	100.0 mL distilled water
Solution 2	50.0 mL 2.05M NH₃	See Note, p. 114

CALCULATIONS

Study carefully the calculations for NH₄Br at the end of this section.

1. Find the temperature change ΔT from the temperature-time plots by extrapolating the plotted data to the time of mixing, reading the initial and final temperatures, and taking the temperature change for each determination. See figure 3 for an example. Use good graphing technique. Data should take up about 2/3rd of the paper. The extrapolation should not run off the top or the bottom of the page. Graph should have a title and axes should be labeled. Origin does *not* need to be (0, 0).

2. Use the calorimeter constant, C_c, of 15 J/K for the calculation.

3. Use the data from Part A and the calorimeter constant to calculate the enthalpy of neutralization of NH₃ and the assigned acid.

4. Use the data from Part B to compute the enthalpy of solution of the ammonium salt.

5. Use the results for the enthalpy of neutralization and the enthalpy of dissolution, along with the literature values of ΔH_f° for NH₃(aq) and the assigned acid (given in Table 3) to compute the enthalpy of formation of the ammonium salt.

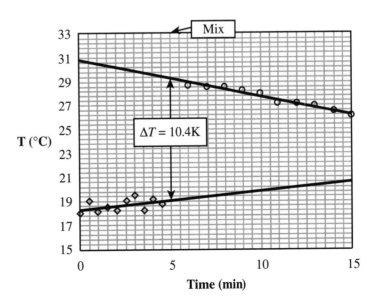

FIGURE 3. *Plot of temperature-time data showing determination of* ΔT.

Typical Calculations for NH₄Br

Part A

$$\Delta T = 11.9 \text{ K}, \quad d = 1.018 \text{ g/mL} \quad C = 3.91 \text{ J/gK}, \quad \text{M.M.} = 97.8 \text{ g/mol}$$

$$\text{Mass of solution} = 100 \text{ mL} \times 1.018 \text{ g/mL} = 101.8 \text{ g}$$

$$C_S \text{ (heat capacity of the solution)} = 101.8 \text{ g} \times 3.91 \text{ J/g K} = 398 \text{ J/K}$$

$$\Delta H_r = -(C_S + C_c)\Delta T = -(398 \text{ J/K} + 15.0 \text{ J/K})11.9 \text{ K} = -4915 \text{ J}$$

$$= -4.915 \text{ kJ}$$

Students mixed 50 mL of 2.00M acid with 50.00 mL of 2.05M base, i.e., 0.100 mol of the reactants were used. Therefore, the measured ΔH_r is for 0.100 mol of product formed.

$$\text{molar } \Delta H_r = -4.915 \text{ kJ}/ 0.100 \text{ mol} = -49.2 \text{ kJ/mol}$$

Part B

TABLE 2
MOLECULAR WEIGHTS, HEAT CAPACITIES, AND DENSITIES OF SOLUTIONS

Solute	Molecular wt. $g\ mol^{-1}$	Concentration $mol\ L^{-1}$	Sp. heat capacity[a] $J\ g^{-1}\ K^{-1}$	density[a] $g\ mL^{-1}$
NH_4Cl	53.49	1.00	3.93	1.013
NH_4NO_3	80.04	1.00	3.90	1.029
H_2O	18.02	—	4.18	0.997

[a]At 25°C

TABLE 3
ENTHALPIES OF FORMATION AT 25 °C[a]

Reaction	$\Delta H_f^o\ kJ\ mole^{-1}$
$\frac{1}{2}N_2(g) + \frac{3}{2}H_2(g) + \xrightarrow{H_2O} NH_3(aq, 2.0M)$	−80.7
$\frac{1}{2}H_2(g) + \frac{1}{2}Cl_2(g) \xrightarrow{H_2O} HCl(aq, 2.0M)$	−164.7
$\frac{1}{2}H_2(g) + \frac{1}{2}N_2(g) + \frac{3}{2}O_2(g) \xrightarrow{H_2O} HNO_3 (aq, 2.0M)$	−206.0
$H_2(g) + \frac{1}{2}O_2(g) \rightarrow H_2O(l)$	−285.8

[a] "Selected Values of Chemical Thermodynamic Properties," Circular 500, National Bureau of Standards, Washington, D.C., 1961

From Data Sheet 1, the number of moles of the product formed is 0.100 mol. Therefore, the mass of the NH_4Br used to measure the enthalpy of dissolution is: 97.9 g/mol $\times$ 0.100 mol = 9.79 g.

$$\Delta T = -3.5 \text{ K} \quad \text{Mass of solution} = 100 \text{ mL } H_2O \times 1 \text{ g/mL} + 9.79 \text{ g } NH_4Br = 109.79 \text{ g}$$

$$C_S = 109.79 \text{ g} \times 3.91 \text{ J/g K} = 429 \text{ J/K}$$

$$\Delta H_r = -(429 \text{ J/K} + 15.0 \text{ J/K})(-3.5 \text{ K}) = 1.6 \times 10^3 \text{ J}$$

(handwritten left margin:)
−80.7 kJ/mol
−206.0 kJ/mol
−45.79 kJ/mol
−24.89 kJ/mol
−357.4 kJ/mol

$$= 1.6 \text{ kJ}$$

molar ΔH_r = 16 kJ/mol

Calculation of ΔH_f^0 of $NH_4Br(s)$

$$\frac{1}{2} N_2(g) + \frac{3}{2} H_2(g) \xrightarrow{\;H_2O\;} NH_3(aq) \qquad \Delta H_1 = -80.7 \text{ kJ/mol}$$

$$\frac{1}{2} H_2(g) + \frac{1}{2} Br_2(l) + \xrightarrow{\;H_2O\;} HBr(aq) \qquad \Delta H_2 = -174 \text{ kJ/mol}$$

$$NH_3(aq) + HBr(aq) \longrightarrow NH_4Br(aq) \qquad \Delta H_3 = -42.9 \text{ kJ/mol (from Part A)}$$

$$NH_4Br(aq) \xrightarrow{\;-H_2O\;} NH_4Br(s) \qquad \Delta H4 = -16 \text{ kJ/mol (from Part B)*}$$

Sum up all four equations

$$\frac{1}{2} N_2(g) + \frac{3}{2} H_2(g) + \frac{1}{2} Br_2(l) \rightarrow NH_4Br(s)$$

$$\Delta H_f^0[NH_4Br(s)] = \Delta H_1 + \Delta H_2 + \Delta H_3 + \Delta H_4 = -80.7 + (-174) + (-42.9) + (-16) = -320 \text{ kJ/mol}$$

(handwritten upper right:)
$$\frac{1}{2} N_2(g) + \frac{3}{2} H_2(g) \rightarrow NH_3(aq)$$
$$\frac{1}{2} H_2(g) + \frac{1}{2} N_2(g) + \frac{3}{2} O_2(g) \rightarrow HNO_3(aq)$$
$$NH_3(aq) + HNO_3(aq) \rightarrow NH_4NO_3(aq)$$
$$NH_4NO_3(s) \rightleftharpoons NH_4NO_3(aq)$$
$$\overline{N_2(g) + 2H_2(g) + \frac{3}{2}O_2(g) \rightarrow NH_4NO_3(s)}$$

(handwritten bottom:)
$$HNO_3 + NH_3 \longrightarrow \boxed{NH_4NO_3}$$
$$2M \qquad 2.05 M \qquad 0.100 \text{ mol} \times \left(\frac{80.043 \text{ g}}{1 \text{ mol}}\right)$$
$$.050 L \qquad .050 L$$
$$\boxed{0.100 \text{ mol}} \quad 0.1025 \text{ mol} \qquad 8.0043 \text{ g}$$
$$0.0025$$

_____
* The sign is different from Part B; since the equation is reversed.

VOLUMETRIC ANALYSIS: AN ACID-BASE TITRATION

INTRODUCTION

In this experiment you will learn how to use volumetric glassware: a buret, a volumetric flask, and a pipet. Prior to use, you will have to clean and condition all the glassware. Your first task will be to determine the molarity of the NaOH solution using the solid primary standard, oxalic acid dihydrate $H_2C_2O_4 \bullet 2H_2O$. A *primary standard* is any highly stable substance whose purity is extremely high from which we can determine the concentration of another solution. Ideally, a primary standard will have a high molecular weight. Oxalic acid is a diprotic acid and will neutralize the base, sodium hydroxide. The stoichiometric relationship between the acid and the base can be used to determine the molarity of the base. This relationship can be calculated from the grams of acid required to react with a specific volume of base.

Once the base solution, NaOH, has been standardized and its molarity determined, it will be used to titrate a hydrochloric acid solution that you will dilute from a stock solution. The base is acting as a *secondary standard* because we are using the concentration of the base as determined from the primary standard to find the concentration of the acid.

A titration is the use of a buret to quantitatively deliver a known amount of a solution of known concentration as it reacts with a known amount of solution of unknown concentration. When the reaction between the two solutions is complete, it is called the *stoichiometric point* or *equivalence point*. At this point the two reactants are chemically equivalent. *The stoichiometric relationship of the reaction determines this point.* The solution whose concentration is known is called the *titrant,* the solution whose concentration is being analyzed is called the *analyte.* Not all chemical reactions take place in colored solutions; therefore, it is necessary to use an indicator to denote the end point. An *indicator* is usually an organic dye that changes color when the reaction is complete. This color change indicates to the observer that the reaction is complete and is called the *end point.* Normally an indicator is one color in an acid and a different color in a base. The indicator may not change color *exactly* at the equivalence point. The difference between the end point and the equivalence point is called the *titration error.* If we have picked the correct indicator, the titration error will be small enough to ignore. When an acid and a base react, water and salt are the products of this "neutralization" reaction. Near the end point of a reaction, the number of hydrogen ions and the number of hydroxide ions present reach equilibrium, mol H^+ = mol OH^-. At this equivalence point, the solution is neutral: neither acidic nor basic. Once the end point is reached, the addition of either more acid or base will cause the indicator to change color thus identifying the end point to the chemist. The indicator you will be using is phenolphthalein which is colorless in acid and turns pink in base.

Adapted from *General Chemistry Laboratory Manual.* Fourth Edition by D.L. Stevens. Copyright © 2004 by Kendall/Hunt Publishing Company. Reprinted by permission.

For a titration to be useful, the reaction must be complete, the reaction should be fast, there should be no side reactions, and there must be a method of detecting the end point. The reaction we will use to standardize the base is:

$$H_2C_2O_4 \bullet 2H_2O(aq) + 2NaOH(aq) \rightarrow Na_2C_2O_4(aq) + 4H_2O(l)$$

An old measure of concentration, still used in the laboratory is **_normality._** With HCl and NaOH, the neutralization reaction is $HCl(aq) + NaOH(aq) \rightarrow NaCl(aq) + H_2O(l)$. One mole of acid neutralizes one mole of base. They are equivalent. Now consider sulfuric acid. The reaction is

$$H_2SO_4(aq) + 2NaOH(aq) \rightarrow Na_2SO_4(aq) + 2H_2O(l)$$

1 mol	2 mol
2 equiv.	2 equiv.

Two moles of NaOH are required to completely neutralize 1 mole of H_2SO_4. Thus, sulfuric acid has 2 equivalents per mole.

$$\# \ of \ equivalents = \frac{mass}{equivalent \ mass}$$

$$equivalent \ mass = \frac{molar \ mass}{n}, \ \text{where } n \text{ is the number of transferable protons.}$$

This is useful because in the lab, we detect when the H^+, *or its equivalent* is used up. Thus, we are interested in the concentration of H^+, *or its equivalent.* For bases, we are interested in $[OH^-]$, *or its equivalent.* Just as molarity, M, is the concentration of a compound or ion measured as moles of solute per liter of solution, **_normality_** is equivalents per liter of solution.

$$M_{acid} = \frac{n_{acid}}{V_{acid}} \qquad N_{acid} = \frac{\# \ equiv. \ acid}{V_{acid}}$$

The advantage of normality is that at the equivalence point, the number of moles of H^+, *or its equivalent,* is equal to the number of moles of OH^-, *or its equivalence.* Thus, $N_A V_A = N_B V_B$ regardless of whether the acid is mono-, di-, or triprotic.

LABORATORY SAFETY

In this experiment several safety precautions need to be observed, they are:

Approved safety goggles & appropriate lab clothes must be worn at all times while in the laboratory.

The buret, pipet and volumetric flask, being elongated pieces of equipment, need to be handled with care. They can be accidentally broken by undue stress or by dropping it. A buret must be physically *held* while being inserted or removed from the buret clamp. If not held properly, gravity will control its behavior.

The standard base solution, sodium hydroxide, NaOH, even though a dilute solution can be very caustic (tissue destroying). If you physically come in contact with NaOH, thoroughly wash it off. A good precaution is to always wash your hands after working with NaOH. You may or may not have some NaOH on your hand. If you accidentally rubbed your eye with a hand contaminated with NaOH, it could get into your eye and cause blindness. Always wash your hands after working with NaOH.

The hydrochloric acid solution being used in this experiment, though dilute, can irritate your skin. If you come in contact with it, immediately wash the affected area several minutes with plenty of water. Acid spills should be contained. Powdered $NaHCO_3$ can be spread over the spill to neutralize it.

If at anytime during this experiment, you find yourself excessively itching any area of your arms or hands, you probably have a small amount of acid on that particular area. Immediately wash the area with plenty of water for several minutes.

Oxalic acid is poisonous. It is common in rhubarb leaves. Avoid physical contact with it and when done using it, wash your hands and arms thoroughly with soap and water to remove any possible traces.

Please do not be wasteful with the reagents, as chemical wastes have to be properly disposed of in special landfills, by incineration, or by purification. By reducing the amount of waste you generate, you reduce the cost of disposing of them. **When you are finished, save the acid and base as instructed. You will use them again for the next lab.**

THE PREPARATION OF THE HCl

Volumetric glassware comes in two types: TC glassware is calibrated *to contain* a precise amount; TD is calibrated *to deliver* a precise volume. The volumetric flask contains precisely the volume printed on it when it is filled to the calibration line etched in its neck. Obtain a 250-mL volumetric flask. Make sure it is clean. Rinse it several times with distilled water. Put some distilled water in the flask (the water level should be about 1 cm deep). Take about 30 mL of the stock 6M HCl solution to your bench. Measure 25 mL in a graduated cylinder. Carefully pour the acid into the flask.

Slowly add water to the flask, mixing periodically by swirling the flask. When the solution reaches the neck, put the cap on the flask and invert it several times. Fill the flask until the bottom of the meniscus just reaches the etched line on the flask. Add the last amount of water one drop at a time. **DO NOT OVERFILL THE FLASK! ONCE LIQUID HAS BEEN ADDED TO THE SOLUTION, IT CANNOT BE TAKEN OUT. POURING OUT EXCESS WATER WILL RESULT IN A SOLUTION OF THE WRONG CONCENTRATION.**

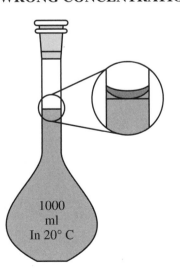

Cap the flask and invert 5–6 times. The volume within the flask will be 250.00 mL.

1000
ml
In 20° C

THE PREPARATION AND USE OF A BURET

Obtain approximately 200 mL of 1M NaOH. Get your solution early. If you have to come back to get more, you will be penalized and you might not get base of exactly the same concentration. Keep the NaOH covered. Sodium hydroxide solutions absorb CO_2 from the atmosphere very slowly. Over a period of several hours to days, the concentration can change due to the reaction:

$$2NaOH(aq) + CO_2(g) \rightarrow Na_2CO_3(aq) + H_2O(l).$$

Clean a 50.00 mL buret with soap and water. Rinse well, first with tap water, then with distilled water. To rinse a buret, pour between 5–15 mL of water in the barrel of the buret. Let approximately 2 mL run out the tip. Dump the rest out the back of the buret, twisting as you go to wet the entire barrel. **NEVER LET ALL THE LIQUID RUN OUT THE TIP OF A BURET!** This can leave air bubbles that can impede the function of the buret or lead to an incorrect volume being dispensed. If liquid does not readily run out the tip, tell your instructor. Rinse the buret 3–4 times with distilled water, then with NaOH. Using a buret clamp, mount the 50.00 mL buret on a ring stand. Fill the 50.00 mL buret with 1 M NaOH solution. Never adjust the volume of a buret to an exact amount (e.g., 0.00). This is a waste of time.

A buret is a TD piece of glassware. It has graduations marked on it. Zero is at the top and 50 is at the bottom. The smallest graduation is 0.10 mL. Because you can estimate between lines, a buret can be read to ±0.02mL. To read a buret, hold a white piece of paper with a dark line on it behind the buret. The dark line will cast a shadow on the bottom of the meniscus, making it easier to read. Look straight at the meniscus to avoid parallax. The buret can be raised or lowered to make it easier to read. The reading on the buret pictured below is 12.68 mL. Never go past the 50 mL mark during a titration. If you have gotten to 49 mL without reaching the end point, **STOP!** Make sure you have indicator in your mixture. Record the volume, refill the buret and record a new starting volume. To find the volume dispensed, you will have to add the two volumes together.

$$V = (V_{f.1} - V_{i.1}) + (V_{f.2} - V_{i.2})$$

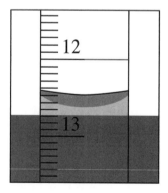

THE STANDARDIZATION OF A BASE SOLUTION, NaOH

Weigh 0.8–1.0 grams of solid oxalic acid to the nearest 1 mg. *Note:* molar mass of oxalic acid is 126.07g/mol because you must include the mass of the water of hydration. Transfer all the oxalic acid to a clean 125-mL Erlenmeyer flask (it doesn't have to be dry). Add approximately 40 mL of distilled water to the flask. Swirl the flask until all the oxalic acid is dissolved. Add two drops of phenolphthalein.

Titrate the flask with the 1 M NaOH solution as follows. Record the initial volume in the buret. Open the stopcock on the buret and let NaOH flow into the flask, swirling the flask at the same time. When you start to see pink flashes in the solution, start adding the base more slowly. Eventually, you will add the base one drop at a time. Occasionally, rinse the tip of the buret and the sides of the flask with a small amount of distilled water. You have reached the end point when the entire solution remains light pink for 30s. Record the final NaOH buret reading. Perform a second titration and calculate the molarity (M) of the NaOH solution. If the concentrations do not agree to within 0.05M, you may need to perform a 3rd titration.

STANDARDIZING THE HCl SOLUTION

Use of a Pipet

A pipet is a TD piece of volumetric glassware. Solution is drawn into it using a rubber bulb. When the solution just touches the calibration line, the amount of liquid that will run out of it will be exactly the volume printed on the pipet. You will be using a 25-mL pipet. Properly used, it will deliver 25.00 mL of solution.

Solutions should be drawn in from small beakers, *never from volumetric flasks*. Also, you never use a pipet directly from a stock solution. You can contaminate the entire supply.

To use a pipet, squeeze the rubber bulb. Just touch the end of the bulb to the lip of the piper. ***DO NOT FORCE THE BULB OVER THE TIP OF THE PIPET!*** With slight upward pressure from the hand holding the pipet, put the tip in the liquid to be drawn in. The hand holding the pipet should be as close to the bulb as you can get and still control the pipet. Slowly release the bulb, sucking in the liquid. When the level rises over the etched line, push the bulb away with the index finger of the hand holding the bulb and cover the hole at the top of the pipet. Learn to use your index finger. Beginning students are tempted to use their thumbs. The thumb is less sensitive and gives you less control. Put the tip down on the bottom of the beaker, *gently*. By arching your finger, or even lifting it off slightly, allow the level to drop until the meniscus just touches the line. Clamp your finger down. Move the tip to the container into which you will put the solution and let it flow. **DO NOT BLOW OUT THE LAST DROPS. THE PIPET HAS BEEN CALIBRATED TO** *DELIVER,* **NOT** *CONTAIN* **25 mL.**

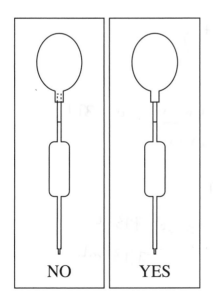

NO   YES

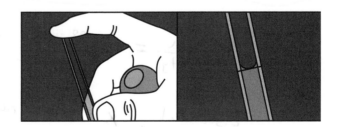

You may need to practice with water several times. Good pipetting technique is a valuable lab tool. Many analyses require pipetting solutions.

Cleaning and Conditioning a Pipet

Clean the pipet by running lots of tap water through it. You may need to use a small amount of soapy water to clean it. Rinse the pipet several times, first with tap water, then with distilled water.

Pour some of your acid into a 100-mL beaker. Use the bulb to fill the pipet $1/3^{rd}$ full. Holding the pipet parallel to the floor, twist the pipet so that the solution coats the sides of the pipet. Let the solution run down the drain with lots of running water. Repeat 2 or 3 times.

Standardizing the Acid

Refill the NaOH buret with the standard 1M NaOH solution, if needed. Pipet 25.00 mL of your acid into a clean 125-mL Erlenmeyer flask. They do not have to be dry. Add 2 drops of phenolphthalein indicator to the Erlenmeyer flask. Titrate the flask with the standard 1M NaOH solution. Record the initial and the final NaOH buret reading for the titration. Do a second titration and calculate the concentration of your acid. If the concentrations differ by more than 0.05M, you may need to perform a third run.

$$H_2C_2O_4 \cdot 2H_2O \qquad \frac{0.2500 \text{ mol}}{1 L} \times 0.03856 \text{ L} = .009640 \text{ mol} \quad \text{oxalic A} \atop NaOH$$

$$H_2C_2O_4 \cdot 2H_2O + 2NaOH \rightleftharpoons Na_2C_2O_4 \cdot 2H_2O + 2H_2O$$

$$0.004820 \qquad 0.009640$$

$$.004820 \text{ mol} \times \frac{126.07 g}{1 \text{ mol}} = 0.6077 \text{ g}$$

$$.250 \text{ g} \times \frac{1 \text{ mol}}{126.07 g} = 0.00198 \text{ mol} \times \frac{2 \text{ mol NaOH}}{1 \text{ mol oxA}} = 0.00397 \text{ mol NaOH}$$

$$\frac{0.500 \text{ mol}}{1 L} \times xL = 0.00397$$

$$xL = \frac{0.00397}{.500} = .00793 \text{ L}$$

$$7.93 \text{ mL}$$

EVALUATING COMMERCIAL ANTACIDS

INTRODUCTION

All too often, labs are dry, canned exercises with very little connection to real world chemistry. This experiment is designed to show you an actual application of the chemistry you have been learning. Over-the-counter antacid preparations relieve heartburn or acid indigestion. These symptoms are caused by excess stomach acid, which, is mostly hydrochloric acid (HCl). Antacids contain basic substances that either neutralize the excess acid or act in the stomach. The types of bases commonly used in antacids include metallic hydroxides, metallic carbonates, or hydrogen carbonates. Antacids should not be confused by a new class of drugs like Prevacid® that block the production of acid in the stomach.

Some reasonable questions to ask are: how well do antacids work; is there a difference in effectiveness from one brand to another; which antacid is most cost effective? To answer these questions, we need an experimental system to measure the neutralizing power of antacids. We model the stomach acid with the HCl solution you standardized last lab. We will neutralize it with the antacid. The net ionic equation for the neutralization of magnesium hydroxide, $Mg(OH)_2$, a typical metallic hydroxide used in antacids, is shown in equation 1.

$$Mg(OH)_2(s) + 2H^+(aq) \rightarrow Mg^{2+}(aq) + 2H_2O(l) \hspace{2cm} \text{(Eq. 1)}$$

In this reaction, 2 moles of hydrogen ions react with 2 moles of hydroxide ions (OH^-). Thus, 1 mole of hydroxide ions neutralizes 1 mole of H^+ ions.

Many antacids are metallic carbonates, such as calcium carbonate ($CaCO_3$). The net ionic equation for the neutralization of $CaCO_3$, is shown in equation 2.

$$CaCO_3(s) + 2H^+(aq) \rightarrow Ca^{2+}(aq) + CO_2(g) + 2H_2O(l) \hspace{2cm} \text{(Eq. 2)}$$

In the reaction in equation 2, 1 mole of carbonate ions (CO_3^{2-}) neutralizes 2 moles of H^+ ions. Note that this produces a gas. A similar reaction accounts for the fizzing of antacids like Alkaseltzer® that contains $NaHCO_3$.

Our measure of effectiveness of antacid preparations will be the number of moles of HCl that will react with a known mass of antacid. To accomplish this, we can use an analytical procedure called back-titration. In a back-titration, we add more standardized HCl solution to the antacid tablet than the tablet can neutralize. Then, we titrate the un-neutralized, or excess, HCl with standardized sodium hydroxide solution (NaOH). The net ionic equation for the back-titration reaction is shown in equation 3.

$$OH^-(aq) + H^+(aq) \rightarrow H_2O(l) \hspace{2cm} \text{(Eq. 3)}$$

The hydroxide comes from the standardized solution, the acid is the excess. This has a number of advantages. Some of the bases used in antacids are only partially soluble. Over titrating the base eliminates the solubility problem. Secondly, the end-point for a strong acid-strong base titration is sharper than with the bases in the antacids, which are intentionally weaker or slower to dissolve to be easier on the stomach.

The concentration of the acid allows us to determine the number of moles of HCl we used to "over neutralize" the antacid. The reaction in equation 3 allows us to determine the number of moles of HCl left; the excess HCl added. By subtracting the number of moles of HCl we titrate from the number of moles of HCl we added initially, we can determine the number of moles of HCl neutralized by the antacid.

Our scheme can be summarized by equations 4 and 5 below.

$$\text{Antacid(s)} + \text{H}^+(\text{aq, from std. HCl}) \rightarrow \text{H}^+(\text{aq}) + \text{neutral ion(aq)} \qquad \text{(Eq. 4)}$$

$$\text{H}^+(\text{aq, excess from 4}) + \text{OH}^-(\text{aq, from std. NaOH}) \rightarrow \text{H}_2\text{O}(l) \qquad \text{(Eq. 5)}$$

We calculate N, the number of moles of HCl neutralized by a tablet, from equation 6, where A is the number of moles of HCl initially added to the tablet, and B is the amount of NaOH required for the back-titration.

$$N = A - B \qquad \text{(Eq. 6)}$$

Equation 7 is used to calculate A, the number of moles of acid added, where V_A is the volume of acid added to a tablet, and M_A is the concentration of the acid used.

$$A = V_A \times M_A \qquad \text{(Eq. 7)}$$

The number of moles of NaOH required for the back-titration, B, is calculated from equation 8. V_B is the volume of standardized base required for the back-titration of the excess acid, and M_B is the concentration of that base. Equation 3 shows that the number of moles of NaOH required equals the number of moles of excess acid. Diprotic acids, like H_2SO_4, produce 2 moles of H^+ per mole of compound. Thus, if we define M_A to be the concentration of H^+, rather than the acid itself, $M_A = 2 \times M_{\text{acid}}$. This definition is the rationale behind normality.

$$B = V_B \times M_B \qquad \text{(Eq. 8)}$$

The effectiveness of an antacid is measured by how much acid that formulation can neutralize. We can measure the effectiveness on a number of bases: by tablet, by mole, by volume, by mass, or by cost. The average shopper will not care about the molar effectiveness of the antacid, and variations in the manufacturing process might mean the volume and amount of active ingredient could vary from tablet to tablet. The two most useful comparisons are the mass effectiveness and the cost effectiveness, which we will call E and C. The mass effectiveness of an antacid is the number of moles of HCl neutralized per gram of tablet, using equation 9, where N is the number of moles of HCl neutralized by a tablet, and W is the mass of the tablet in grams.

$$E = \frac{N}{W} \qquad \text{(Eq. 9)}$$

We can calculate C the cost effectiveness of a tablet, which is the number of moles of HCl neutralized per one cent using equation 10. N is the average number of moles of HCl neutralized per tablet, and P is the cost of one tablet in cents. Cost effectiveness can be measured per dollar or per penny. We will be measuring it per penny. We use the average number of moles $\overline{N}$ to account for any variations in the manufacturing process. You will be analyzing 2 tablets. To get a better estimate of $\overline{N}$, you should really analyze at least 10 tablets.

$$C = \frac{\overline{N}}{P}$$

(Eq. 10)

PROCEDURE

CHEMICAL ALERT: Both 0.6M hydrochloric acid and 1.0M sodium hydroxide are toxic and corrosive. CAUTION: **Wear departmentally approved eye protection while performing this experiment.**

Note: Avoid handling the antacid tablets with your bare fingers as much as you can. Any oil or dirt transferred from your fingers to a tablet will adversely affect the outcome of this experiment.

Obtain two antacid tablets of the same brand. The stockroom will provide these unknowns in a plastic bag. Record on your data sheet the unknown code of your antacid. The cost per 25 tablets will be posted around the laboratory.

I. Transferring HCl with a Volumetric Pipet

On your data sheet, record the molarity of the standardized HCl solution you saved from last week's experiment.

Rinse a clean 25-mL pipet by drawing about 5 mL of distilled water into the pipet (pronounced pipe'-et), using a rubber bulb. **Remember, never force the bulb over the tip of the pipet.** The mouth of the bulb should abut the top of the pipet. Quickly disconnect the bulb, and place your index finger over the top of the pipet to prevent the water from draining. Learn to use your finger rather than your thumb. You have more control with your finger than with your thumb. Hold the pipet in a nearly horizontal position. Rotate the pipet to allow the water to contact all interior surfaces. Remove your index finger briefly from the top of the pipet during this process, in order to allow the water to enter the upper stem of the pipet. Allow the water to drain from the pipet through the tip.

In a similar fashion, rinse the pipet with approximately 5 mL of your standardized HCl solution. Discard the rinse solution into the sink with running water. Repeat the rinsing procedure twice, using a 5-mL portion of your HCl solution each time.

Note: After you dispense a solution from a pipet, a small amount of liquid will remain in the pipet tip. **Do not** blow this liquid out of the pipet. A 25-mL pipet is calibrated to deliver (TD) 25.00 mL of solution *excluding* the small amount remaining in the pipet tip.

Carefully pipet 25.00 mL of your standardized HCl solution into a clean 125-mL Erlenmeyer flask. As you release the solution into the flask, hold the pipet tip against the inside wall of the flask. Allow the solution to flow down the inside wall of the flask, in order to prevent splattering. After you deliver the solution from the pipet, allow the pipet to drain while you continue holding the tip against the inside wall of the flask for an additional 15 s to ensure all the solution drains from the pipet into the flask. You may also rise the walls of the flask with a small amount of distilled (DI) water to ensure that all the solution gets to the bottom of the flask.

II. Preparing the Buret

Record on your data sheet the molarity of the NaOH solution that you standardized last week. If your standardization last week was off by a small amount you should use your value. If your standardization was too far off, your instructor will provide you with a value. Although this seems like a "double penalty" for getting the wrong value, in reality the errors in titrations tend to cancel out. Thus, if you tend to go past the end point in your titrations and got poor values for your standard solutions last week then you will likely do the same thing this week. Using your "bad" values will give more accurate results than if we corrected them. This will not be true if your calculated concentrations are too far off.

Rinse your buret (pronounced beur'-et) thoroughly with tap water. Add approximately 10 mL of distilled water to a clean 50-mL buret. Hold the buret in a nearly horizontal position. Rotate the buret to allow the water to contact all interior surfaces of the buret. Drain about 1 mL of the water through the buret tip. Pour the remaining water out the top of the buret. **Never run a buret dry. If you let all the liquid run out through the tip, you run the risk of getting an air bubble trapped in the stopcock. Bubbles can impede the proper functioning of the buret or come out in the middle of a titration making your volume inaccurate.** Repeat this procedure twice, using a 10-mL portion of water each time.

In similar fashion, rinse your buret with three 5-mL portions of the standardized NaOH solution. Drain the rinse solutions through the buret tip and out the top into the sink with running water.

Note: When you fill your buret in the following step, the buret tip must be completely filled with NaOH solution. Remove any air bubbles in the buret tip before you begin the titration. If the bubbles are not removed, the exact volume of solution used in the titration will be in question.

Clamp the buret onto a ring stand. Close the stopcock. Fill the buret with your standardized NaOH solution. Remove any air bubbles trapped in the tip of the buret by allowing the solution to flow through the stopcock into your "Discarded Solutions" container until the bubbles are forced out of the tip. If you still see an air bubble, inform your instructor. Lower the solution level in the buret below the 0.00-mL mark. Don't waste time adjusting the starting value to exactly 0.00. Too many readings of 0.00 either mean the data is manufactured (fraudulent) or the student is wasting time. In either case points will be deducted from your grade. Read the buret to the nearest 0.02 mL. Record this initial buret reading on your data sheet.

III. Analyzing an Antacid Preparation

Note: In the next step, you will add a portion of distilled or deionized water to the acid in the flask, in order to increase the solution volume for the titration. The number of moles of acid in the solution will be the same after the addition of water as it was before. Thus, it is not necessary to know the new volume of the solution or its concentration.

Using a graduated cylinder, add approximately 40 mL of distilled water to the Erlenmeyer flask into which you pipetted the 25.00 mL of HCl.

Place one of your antacid tablets on a piece of weighing paper. Weigh the tablet and paper to the nearest milligram (0.001 g). Record this mass on your data sheet. Place the wax paper on the same balance. Determine its mass to the nearest milligram. Record this mass on your data sheet.

Note: In the next sequence of steps, you will boil the antacid-HCl solution to expel the dissolved CO_2. Any CO_2 present during the subsequent titration will interfere with the determination of the end point.

Put the tablet in the plastic bag it came in and crush it. Add the weighed tablet to the HCl solution in the Erlenmeyer flask. Heat the flask very gently. Do not let the solution boil over. Remove the burner if bubbles start to form too quickly. After boiling for 2–3 minutes, remove the flask carefully and place it on a paper towel. Wait for the solution to cool. Add 10 drops of bromothymol blue indicator. Mix the solution thoroughly by swirling the flask and contents. Do not be concerned if some solid particles remain suspended in the mixture after boiling. Many antacids contain inert materials called binders that keep the tablet held together until it reaches the stomach. These materials are essentially glues and will not dissolve under these conditions.

If the solution is not yellow then the table will neutralize more acid than you have added. Add another 25 mL of standardized HCl, repeat the heating, and add another 10 drops of indicator.

Touch the buret tip to the inside wall of the "Discarded Solutions" beaker to remove any NaOH solution that may be clinging to the tip. Adjust the buret so that the tip is 2–3 cm below the rim of the mouth of the Erlenmeyer flask containing the sample, as shown in Figure 1.

As the reaction approaches the titration end point, you will notice momentary blue-green flashes in the area where the NaOH solution enters the solution in the flask. When the reaction is very close to the end point, these blue-green areas will grow larger, taking longer to revert to yellow. At this point, you should add the NaOH solution dropwise. The titration is complete when the reaction mixture in the flask becomes green and remains green for 15 s after thorough mixing. Note the volume and add one more drop. The solutions should turn sky blue just past the end point. It is possible that you will go right through the end point, going straight from yellow to blue. The titration error will be so small as to be negligible. The volume of 1 drop will not alter your results either way.

Swirl the Erlenmeyer flask with one hand. Control the stopcock of the buret with your other hand. Begin to titrate the excess HCl in the solution by slowly adding NaOH solution from the buret. Continue to swirl the flask throughout the titration. Near the end of the titration, add NaOH solution in smaller and smaller increments until you are making dropwise additions. When you reach the end point, read the NaOH level in the buret to the nearest 0.02 mL. Record this final buret reading on your data sheet. Experienced chemists can actually add half a drop at a time, but this requires practice and a steady hand. Frequent rinsings with distilled water to get all the base into the reaction mixture may be needed.

Do a second determination, using a second tablet, 25.00-mL HCl sample, and Erlenmeyer flask. Your second flask must be clean, but not dry. Record all your data in the appropriate columns of your data sheet.

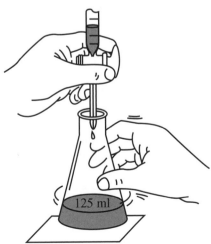

FIGURE 1. *Position of buret tip in mouth of flask.*

You may not need to refill the buret between determinations. Estimate the volume of NaOH solution you will need for a second titration from the volume of titrant used for the first titration. Never let the volume drop below the 50-mL mark; you will have no way to determine how much base you delivered. If enough NaOH solution remains in the buret for the second titration, do not refill the buret. Always record the initial buret reading before beginning a titration.

After completing your last determination, drain the solution from the buret into the "Discarded Solutions" beaker. Rinse the buret several times with distilled water, collecting the rinses in your discard beaker.

Place the contents of the 125-mL flasks into the "Discarded Solutions" beaker. Wash all glassware with tap water and rinse with distilled water. Return the glassware to its proper place.

IV. Treating Discarded Solutions for Disposal

Combine all leftover HCl, NaOH, and discarded solution. Pour the solution into the drain, diluting with a large amount of running water.

CAUTION: Wash your hands thoroughly with soap or detergent before leaving the laboratory.

Calculations

(Do the following calculations and record the results on your data sheet.)

1. Calculate the mass of each antacid tablet [d–e].

2. Calculate the number of moles of HCl added to the tablet in each determination, using equation 5. [b × V_{HCl}]

3. Calculate the number of moles of NaOH required to back-titrate the excess HCl in each determination, using equation 6. [c × i/1000]

4. Calculate the number of moles of HCl neutralized per tablet in each determination, using equation 4 [j–k].

5. Calculate the average number of moles of HCl neutralized per tablet.

6. Calculate the number of moles of HCl neutralized per gram of tablet for each determination, using equation 7. [l/f]

7. Calculate the average number of moles of HCl neutralized per gram of tablet, which is the mass effectiveness of the antacid.

8. Calculate the cost of one tablet of the antacid, using the total cost of the container of antacid and the number of tablets in it when purchased. [a/25]

9. Calculate the cost effectiveness of the antacid, which is the average number of moles of HCl neutralized per $0.01, using equation 8. [o/result of step 8]

SPECTROSCOPY

BACKGROUND INFORMATION

Most reactions do not go to completion; they are equilibria so reversible reactions. In reversible reactions, while reactants are forming products, some of the products are reverting into reactants. We symbolize reversibility in chemical equations using double arrows ($\rightleftharpoons$).

In reversible reactions, the rate of the forward reaction depends on the concentrations of reactants: the rate of the reverse reaction depends on the concentration of the products. At first, the high reactant concentrations cause a rapid forward reaction As the forward reaction proceeds however, reactant concentrations decrease as reactants change into products. Therefore, the forward reaction rate gradually decreases. At the same time, the increasing product concentrations cause the reverse reaction rate to increase. Eventually, the rates of the forward and the reverse reaction rates become equal. Reactants become products just as rapidly as products change back into reactants. There is no further change of either reactants or products. At this point, we say that the reaction has achieved a state of dynamic equilibrium or simply that the reaction is at equilibrium.

Consider the reaction shown in equation 1:

$$I_3^-(aq) + 2S_2O_3^{2-}(aq) \rightleftharpoons 3I^-(aq) + S_4O_6^{2-} \tag{Eq. 1}$$

Regardless of the initial concentrations of reactants or products in this reaction, once the reaction is at equilibrium, the molar concentrations of all three substances will be fixed by the Law of Mass Action as expressed in equation 2, where K is the equilibrium constant for this reaction.

$$K = \frac{[I^-]^3\,[S_4O_6^{2-}]}{[I_3^-]\,[S_2O_3^{2-}]^2} \tag{Eq. 2}$$

We use expressions like that in equation 2 to describe any reaction. We can write a generic reaction as

$$aA + bB \rightleftharpoons cC + dD \tag{Eq. 3}$$

The lower case letters represent the stoichiometric coefficients, while the capital letters represent the compounds. The resulting equilibrium constant is shown in equation 4.

$$K = \frac{[C]^c[D]^d}{[A]^a[B]^b} \tag{Eq. 4}$$

Adapted from *General Chemistry II, Laboratory Manual* by Steven Rowley. Copyright © 2005 by Steven Rowley. Reprinted by permission of Kendall/Hunt Publishing Company.

Equilibrium constants are useful because they allow use to evaluate the compostion of a mixture of reactants and products will be once equilibrium is achieved. Equilibrium constants can range from the very large to the very small. This means that the resulting mixture can be virtually all reactant, nearly equal concentrations of everything or nearly entirely products, or anything in between. Because K depends on temperature, we must measure the concentrations at the same temperature at which K was determined.

FORMING THE ISOTHIOCYANATE ION

Under acidic conditions, the iron (III) ion (Fe^{3+}) will bond with the thiocyanate ion (SCN^-). Rather than bonding with the S, the iron forms a covalent bond with the N to make the $FeNCS^{2+}$ ion, called isothiocyanatoiron (III) ion. The equilibrium is shown in equation 5.

$$Fe^{3+}(aq) + SCN^-(aq) \rightleftharpoons [FeNCS^{2+}](aq) \qquad \text{(Eq. 5)}$$

Fortunately, iron III solutions tend to be pale yellow in color while the $[FeNCS^{2+}]$ ion is a deep, blood red color.

The equilibrium constant for this reaction depends somewhat on the total concentration of all ions (ionic strength), so we will use 0.100M HNO_3 as our solvent for all solutions. This will ensure that the total concentrations of all ions will be roughly equal.

MEASURING THE FENCS^{2+} CONCENTRATIONS

Because the $FeNCS^{2+}(aq)$ ion is intensely colored, we can measure its concentration by measuring the amount of light that passes through the solution. The instrument that measures this is called a spectrophotometer. Light of a specific wavelength is sent through a sample. The intensity of light decrease as some of the photons are absorbed by the sample. This decrease is called the absorbance. The amount of light absorbed is proportional to the concentration and the path length of the light through the sample. The equation that relates concentration is call Beer's Law and is given in equation 6.

$$A_\lambda = \varepsilon c \ell \qquad \text{(Eq. 6)}$$

The constant ε is the molar absorbtivity coefficient, more commonly called the extinction coefficient. Its value depends on the wavelength of light used. The concentration is represented by c. The pathlength ℓ is measured in cm. The subscript λ represents the wavelength at which the measurement was made. Each compound of ion will absorb light to a different extent at each frequency.

The absorbance is related to the amount of light that passes through. The more light that is absorbed by the sample, the less light that makes it through the sample. We call this the transmittance, and is usually expressed as a percent of the intensity of light emitted by the source. This is shown in equation 7.

$$\%T = \frac{I}{I_0} \times 100 = \frac{1}{10^{A_\lambda}} \times 100\% \qquad \text{(Eq. 7)}$$

Transmittance can be measured more precisely than absorbance. If we take the log of both sides of equation 7, we end up with equation 8, which is easier to use than equation 7.

$$A_\lambda = \log\left(\frac{I_0}{I}\right) = 2 - \log(\%T) \qquad \text{(Eq. 8)}$$

To determine the concentration of $FeNCS^{2+}(aq)$ in the mixture, this ion must be the only species to absorb at the wavelength used and we either need to know ε and ℓ or we need a calibration curve. A plot of A vs. concentration, called a **Beer's Law plot,** is the calibration curve. A typical example is shown in figure 1.

PREPARING THE BEER'S LAW PLOT AND MEASURING *K*

To produce a Beer's Law plot for the $FeNCS^{2+}$ ion, we need to collect a set of data relating $\%T$ vs. $[FeNCS^{2+}]$. The problem is that the $FeNCS^{2+}$ ion exists in equilibrium with the Fe^{3+} and SCN^- ions. We need to prepare solutions under conditions that the reaction goes nearly to completion. If we mix relatively high concentrations of Fe^{3+} compared to SCN^-, it will result in almost all the SCN^- to be converted into $FeNCS^{2+}$ ion. Under these conditions, we can approximate the concentration of $FeNCS^{2+}$ with the initial SCN^- concentration. At a wavelength of 447 nm, the only species in our mixture that absorbs light is the $FeNCS^{2+}$ ion.

To measure $K,$ we produce another set of samples. This set will be made under conditions that allow all 3 ions to exist in equilibrium. By measuring $\%T$ at 447 nm, we can use our Beer's Law plot to find the $[FeNCS^{2+}]$ in the sample. Stoichiometry then allows us to determine the equilibrium concentrations of the other ions.

Rather than dissolving small amounts of each compound to make our samples, we will make up stock solutions and then dilute them to make the samples to be analyzed. The initial concentrations can be determined from the concentrations of the stock solutions and the volumes used, as in equation 9.

$$[ion]_0 = \frac{V_{added}[stock\ solution]}{V_{total}} \qquad \text{(Eq. 9)}$$

V_{added} is the volume of the stock solution added to the sample. Once we know the equilibrium concentration of $[FeNCS^{2+}]$, we can find the equilibrium concentrations of the other ions from equations 10 and 11.

$$[Fe^{3+}] = [Fe^{3+}]_0 - [FeNCS^{2+}] \qquad \text{(Eq. 10)}$$

$$[SCN^-] = [SCN^-]_0 - [FeNCS^{2+}] \qquad \text{(Eq. 11)}$$

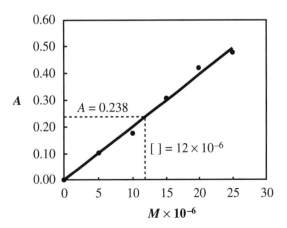

FIGURE 1. *Example of a Beer's Law Plot.*

PROCEDURE

CAUTION: 6M Ammonia is toxic, corrosive and an irritant. Nitric acid is toxic and a strong oxidizing agent. Iron (III) nitrate is toxic, an irritant and an oxidant. Sodium thiocyanate is toxic, corrosive and an oxidant.

You will be working in groups of four for this experiment. You will only work together for the purpose of making solutions and collecting the data. Each student must make his own plot and measure K by him- or herself.

I. Standard Solutions for Beer's Law Plot

1. Label six 50.00-mL volumetric flasks S0 through S5.

2. Pipet 10.00 mL of 0.2M $Fe(NO_3)_3$(aq) solution into each of the flasks. Add the volume of 0.002M NaSCN(aq) indicated in Table 1. There are 2 stock solutions of $Fe(NO_3)_3$ in 0.1M HNO_3: one for the Beer's Law plot (0.2M), and the other for the K measurement (0.002M). Make sure you use the correct solution. S0 is your blank ([FeNCS] = 0). Be careful not to add any NaSCN to this sample. Add 0.1M HNO_3 to bring the volume up to the 50.00 mL line on the flask and mix each solution thoroughly. Record the exact concentrations of the stock solutions on your data sheet.

TABLE 1
STANDARD SOLUTIONS FOR BEER'S LAW PLOT

Solution	$V_{Fe(NO_3)_3}$	V_{NaSCN}	V_{total}
S0	10.00	0.00	50.00
S1	10.00	1.00	50.00
S2	10.00	2.00	50.00
S3	10.00	3.00	50.00
S4	10.00	4.00	50.00
S5	10.00	5.00	50.00

Record the volume of NaSCN added to your data sheet.

II. Analyzing the Standard Solutions

3. Set the wavelength to 447 nm. Adjust the %T to 0.00.

4. Obtain 2 cuvettes. Rinse one with the S0 solution, discarding the solution as instructed. Fill the cuvette with the S0 solution. This will serve as your "blank" or reference sample. Some spectrometers have a single light beam. With these, you adjust the %T manually with the reference sample. Other instruments have dual beam arrangements: 2 light beams and 2 detectors. The reference stays in the light path of one beam and the sample is placed in the other. The instrument then measures the difference in intensity between the two detectors. In either case, use the S0 sample to set 100 %T.

5. Rinse the second cuvette with S1 and discard. Fill the cuvette with S1 and place the cuvette in the spectrometer. Measure and record %T on your data sheet. Discard the solution as instructed.

6. Repeat step 5 for solutions S2–S5. In each case, record the %T on your data sheet.

7. (Optional) If there is sufficient time, repeat your measurements twice more for each solution. If you do this, use the average %T to make your Beer's Law plot.

8. Discard the unused portions of the solutions as instructed. Carefully clean all the glassware in preparation for the next step.

III. Preparation of the Equilibrium Mixtures

9. Obtain six clean 18×150 mm test tubes and label them E0–E5.

10. Pipet 5.00 mL of 0.002M $Fe(NO_3)_3$ solution into each test tube. Pipet the volumes of 0.002M NaSCN listed in Table 2. Be careful! Make sure you use the 0.002M $Fe(NO_3)_3$ and not the 0.2M $Fe(NO_3)_3$. Add the volume of 0.1M HNO_3 listed in Table 2 to bring the total volume to 10.00 mL and mix thoroughly. Record the exact concentrations of the stock solutions on your data sheet.

TABLE 2
STANDARD SOLUTIONS FOR *K* CALCULATION

Solution	$V_{Fe(NO_3)_3}$	V_{NaSCN}	V_{HNO_3}	V_{total}
E0	5.00	0.00	5.00	10.00
E1	5.00	1.00	4.00	10.00
E2	5.00	2.00	3.00	10.00
E3	5.00	3.00	2.00	10.00
E4	5.00	4.00	1.00	10.00
E5	5.00	5.00	0.00	10.00

11. Rinse one of the cuvettes with E0. Discard and fill the cuvette with E0. Adjust %T to 100%. You need a new blank with this set. The solution will scatter some light, which will effect the amount of light that gets through. In addition, although the solvent doesn't absorb appreciably at 447 nm, even a small amount can effect the calculations.

12. Rinse the second cuvette with E1, discard and fill with E1. Measure and record %T on your data sheet.

13. Repeat step 12 with solutions E2–E5.

14. Record the temperature of the equilibrium solutions.

15. Dispose of all solutions as instructed. Clean all glassware, and return any equipement to the stockroom that you checked out. Put the remaining glassware back in your drawer.

CALCULATIONS

I. Beer's Law Plot

1. Calculate the initial concentration of $[SCN^-]_0$ in samples S0–S5 from eq. 9. $[d = c \times b/50]$

2. Calculate $[FeNCS^{2+}]$ by assuming that all the SCN^- reacts. $[e = d]$

3. Calculate the absorbance A_{447} for each sample from eq. 8. $[g = 2 - \log(f)$

4. Prepare a Beer's Law plot by graphing A_{447} on the y-axis and $[FeNCS^{2+}]$ on the x-axis. Draw the best fit straight line through your data. The line should go through the origin.

II. Calculating the Equilibrium Constant

5. Calculate the initial concentrations of Fe^{3+} and SCN^- for solutions E0–E5 from equation 9. $[q = 1 \times h/10.00]$ $[p = k \times i/10.00]$

6. Calculate A_{447} for each sample from eq. 8. $[n = 2 - \log(m)]$

7. Find the equilibrium concentration for each solution E0–E6 by reading it off your graph. Record these in (o).

8. Calculate the equilibrium concentrations of Fe^{3+} and SCN^- from equations 10 and 11. $[s = q - o]$ $[r = p - o]$

9. Calculate K for each solution from equations 4 and 5. $[t = o/(s \times r)]$

10. Calculate the mean K from the results for solutions E1–E5.

PRE-LABORATORY ASSIGNMENT

1. a. A solution gives a reading of 85.6 %*T*. What is the absorbance?

    b. The sample was measured without using a blank. How might that affect the reading?

2. Why is it important that $[FeNCS]^{2+}$ is the only species that absorbs at the frequency we use?

3. Dual beam spectrometers are usually considered more accurate. What problem might make a reading less accurate in a dual beam instrument?

NAME ______ SECTION ______ LOCKER ______ INSTRUCTOR ______

COVER SHEET

EVALUATING THE EQUILIBRIUM CONSTANT FOR THE REACTION OF IRON(III) ION WITH THIOCYANATE ION

Purpose: To determine the equilibrium constant for the formation of the isothiocyanato-iron(III) ion from spectrophotometric data

Procedure: The procedure for this experiment was followed as in this experiment except (list all changes in procedure)

Results:

Exact concentration of 0.2M $Fe(NO_3)_3$ solution (M) ______

Exact concentration of 0.002M $Fe(NO_3)_3$ solution (M) ______

Exact concentration of 0.002M NaSCN solution (M) ______

Mean K ______

Conclusions and Comments:

NAME ______ SECTION ______ LOCKER ______ INSTRUCTOR ______

DATA SHEET 1

Partners: ______ ______

______ ______

Beer's Law Data

a. Exact concentration of 0.2M $Fe(NO_3)_3$ solution (M) ______

b. Exact concentration of 0.002M NaSCN solution (M) ______

	Flask					
	S0	*S1*	*S2*	*S3*	*S4*	*S5*
c. V_{NaSCN} (mL)	______	______	______	______	______	______
d. $[SCN^-]_0$	______	______	______	______	______	______
e. $[FeNCS^{2+}]$	______	______	______	______	______	______
f. *%T*	______	______	______	______	______	______
g. A_{447}	______	______	______	______	______	______

DATA SHEET 2

Data for Determining K

h. Exact concentration of 0.002M $Fe(NO_3)_3$ solution (M) _____

i. Exact concentration of 0.002M NaSCN solution (M) _____

	Flask					
	E0	E1	E2	E3	E4	E5
j. T (°C)	_____	_____	_____	_____	_____	_____
k. V_{NaSCN} (mL)	_____	_____	_____	_____	_____	_____
l. $V_{Fe(NO_3)_3}$	_____	_____	_____	_____	_____	_____
m. %T	_____	_____	_____	_____	_____	_____
n. A_{447}	_____	_____	_____	_____	_____	_____
o. $[FeNCS^{2+}]$	_____	_____	_____	_____	_____	_____
p. $[SCN^-]_0$	_____	_____	_____	_____	_____	_____
q. $[Fe^{3+}]_0$	_____	_____	_____	_____	_____	_____
r. $[SCN^-]$	_____	_____	_____	_____	_____	_____
s. $[Fe^{3+}]$	_____	_____	_____	_____	_____	_____
t. K	_____	_____	_____	_____	_____	_____

Average K _____

POST-LABORATORY QUESTIONS

1. A student measured the calibration curve without using the blank. Later, he measured the blank and found a $\% T$ of 0.002. What effect will this have on the calculated K?

2. Why does a longer pathlength increase the absorbance?

3. How will the equilbirium concentration of $[FeNCS]^{2+}$ change if the pH is raised?

4. What will the equilibrium concentration of Fe^{3+}(aq) be if you mix 1 mol of Fe^{3+} with 1 mol of SCN^- in 1L of solution?

GRAVIMETRIC DETERMINATION OF WATER OF HYDRATION

A hydrate is a substance containing water bound in the molecular form such as alum, $KAl(SO_4)_2 \cdot 12\ H_2O$, and plaster of Paris, $CaSO_4 \cdot 1/2\ H_2O$. The water molecules occupy regular lattice positions and are present in simple stoichiometric ratios. The **theoretical percentage of water** present in a hydrate can be easily calculated by dividing the mass of water per mole of hydrate by the molar mass (including water) of the hydrate (equation 1).

$$\text{Theoretical percent } H_2O = \frac{\text{mass } H_2O \text{ per mol hydrate}}{\text{molar mass}} \times 100 \qquad (1)$$

The water of hydration can often be driven off by mild heating leading to the anhydrous (i.e., without water) form of the substance as illustrated in equation 2. The anhydrous forms of many solids are used as drying agents.

$$CaCl_2 \cdot 2H_2O \xrightarrow{\ heat\ } CaCl_2 + 2\ H_2O(g) \qquad (2)$$

The percentage of water of hydration in a given material can be experimentally determined by heating a weighed sample of the substance to drive off the water and reweighing to determine the mass of water lost. The mass loss divided by the mass of the original sample, multiplied by 100 (Equation 3) will give the percentage of water in the hydrate.

$$\text{Experimental percent } H_2O = \frac{\text{mass loss}}{\text{sample mass}} \times 100 \qquad (3)$$

SAFETY

Light the *Bunsen burner* under the hood so that unburned gas will be carried away by the hood fan.

Handle *hot crucibles* with crucible tongs. Hot porcelain looks the same as cold porcelain.

Always wash your *hands* before leaving the laboratory.

PROCEDURE

Clean two crucibles and lids. Using crucible tongs, place the crucibles with their lids on a clay triangle (at this time do not fully cover crucibles with lids!) and heat intensely with a Bunsen burner for about ten minutes to dry the crucibles. Allow the crucibles to cool for one to two minutes and place them in a desiccator for ten to fifteen minutes or until they reach room temperature. **Using crucible tongs,** transfer the crucible and lid to the balance and weigh accurately. Note which crucible and lid correspond to which mass. Using crucible tongs prevents mass inaccuracies due to contamination of the crucibles and lids with oil and moisture from the skin.

Place approximately one gram of the sample in each crucible and again weigh accurately. Place the crucible covers on the respective crucibles and support each on a clay triangle. Heat with a Bunsen burner using a **low flame** for five minutes to prevent spattering of the sample during dehydration. If spattering occurs, the sample may be lost when the crucible and lid are transferred from the ring to the desiccator. Gradually increase the temperature of the flame until you are using a **medium flame** and continue heating for five minutes. Further increase the temperature of the flame and heat for an additional ten minutes. **CAUTION: DO NOT ALLOW CRUCIBLES OR SAMPLES TO BECOME RED HOT SINCE UNDESIRED SAMPLE DECOMPOSITION MAY OCCUR.** Allow the crucibles to cool slightly and then place them in a desiccator until they reach room temperature. Weigh the samples. Repeat the heating procedure heating at medium heat for about 5 to 10 minutes and again weigh the crucibles when they have cooled. If the mass is not within 0.005 g of the previous mass, repeat the drying procedure until constant mass is attained.

■ Write the test sample number from the vial on your data sheet p. 2-8.

■ Record your results on your data sheet.

Waste Disposal: The anhydrous solid should be disposed of in the solid waste container on the side lab bench labeled "Hydrate Waste." Unused hydrate should be disposed of in the same "Hydrate Waste" container.

Clean-up: Clean and replace your equipment to your student locker. Leave the desiccator on the lab bench at your station. Rinse the vial and cap and place them in the designated container for clean vials. Wipe down your lab bench with a sponge. Wash your hands before leaving the lab.

PRE-LABORATORY QUESTIONS

SHOW SET-UP FOR ALL PROBLEMS

1. Calculate the experimental percent water in $NiC_2O_4 \cdot 2H_2O$ if a student found that a 2.491 g sample of $NiC_2O_4 \cdot 2H_2O$ had a mass of 1.926 g after drying. Report answer using proper number of significant and units.

2. Calculate the theoretical percent water in $NiC_2O_4 \cdot 2H_2O$. Use as many significant figures as reported in the answer to question 1.

3. A student experimentally determined the percent water in two separate samples of a test sample hydrate. The values were 37.6% for sample 1 and 35.3% for sample 2. Calculate the range of percent water for these determinations.

4. Write the balanced equation for the reaction where $Na_2CO_3 \cdot 10H_2O$ is heated to drive off the water of hydration.

NAME ____________________ SECTION __________ LOCKER __________ INSTRUCTOR ____________________

COVER SHEET

GRAVIMETRIC DETERMINATION OF WATER OF HYDRATION/PERCENT WATER IN A HYDRATE

Purpose: To determine the percent water in a hydrated salt.

Procedure: The procedure for this experiment was followed as in this experiment except (list all changes to procedure):

Results:

	trial 1	*trial 2*
% water in hydrate	________	________
mean % water in hydrate	________	

Conclusions and Comments:

**Paste test sample number here: _____

SHOW SET-UP FOR ALL STARRED (*) CALCULATIONS!

	Sample 1	Sample 2
1. Mass of crucible, lid and sample	_____	_____
2. Mass of crucible and lid	_____	_____
3. Mass of sample	_____	_____
4. Mass of crucible, lid and sample		
a. after first drying	_____	_____
b. after second drying	_____	_____
c. after third drying if needed)	_____	_____
5. Mass of water lost	_____	_____
6. Percent of water in sample	_____	_____
7. Average percent water in sample	_____	_____
8. Range of percent water in sample	_____	_____

POST-LABORATORY QUESTIONS

1. Why are crucible tongs used to transfer the cooled crucibles to the balance?

2. How would the observed (measured) mass of the crucible differ from the true mass if the mass were measured while the crucible were still hot? (i.e., the student did not wait for the crucible to cool to room temperature)

3. What is a desiccator? What is its specific purpose in this experiment?

4. Why are you instructed to use the same balance throughout the entire experiment?

5. Why must *everyone* heat, cool, and weigh each sample (at least) twice?